Saving the Oregon Trail

Saving the Oregon Trail
Ezra Meeker's Last Grand Quest

DENNIS M. LARSEN

Washington State University Press
Pullman, Washington

WSU PRESS
WASHINGTON STATE UNIVERSITY

Washington State University Press
PO Box 645910
Pullman, Washington 99164-5910
Phone: 800-354-7360
Email: wsupress@wsu.edu
Website: wsupress.wsu.edu

© 2020 by the Board of Regents of Washington State University
All rights reserved
First printing 2020

Printed and bound in the United States of America on pH neutral, acid-free paper. Reproduction or transmission of material contained in this publication in excess of that permitted by copyright law is prohibited without permission in writing from the publisher.

Library of Congress Cataloging-in-Publication Data

Names: Larsen, Dennis M., 1946- author.
Title: Saving the Oregon Trail : Ezra Meeker's Last Grand Quest / Dennis M. Larsen.
Other titles: Ezra Meeker's last grand quest
Description: Pullman, Washington : Washington State University Press, [2020] | Includes bibliographical references and index.
Identifiers: LCCN 2019038477 | ISBN 9780874223743 (paperback)
Subjects: LCSH: Meeker, Ezra, 1830-1928--Last years. | Pioneers--Washington (State)--Puyallup--Biography. | Oregon National Historic Trail--History. | Historic sites--Conservation and restoration--West (U.S.) | Conservationists--Washington (State)--Puyallup--Biography. | Meeker, Ezra, 1830-1928--Travel. | Overland journeys to the Pacific--History--20th century. | Puyallup (Wash.)--Biography.
Classification: LCC F899.P94 L373 2020 | DDC 978/.02092 [B]--dc23
LC record available at https://lccn.loc.gov/2019038477

On the cover: Meeker with his 1928 Oxmobile in Detroit. *Seattle Museum of History and Industry 1986.5G.1920*

Contents

Foreword by Will Bagley	vii
Preface	xi
Introduction	1
1 Telling History His Way	4
2 A War of Words, a Test of Wills	16
3 The Old Oregon Trail Monument Expedition	23
4 Disaster at the Alaska-Yukon-Pacific Exposition	38
5 Retreat to California	49
6 The Second Old Oregon Trail Expedition, 1910	57
7 The Second Expedition, 1911	78
8 The Second Expedition, 1912	92
9 Interlude	105
10 The Pathfinder Expedition, Part 1: 1915–1916	119
11 The Pathfinder Expedition, Part 2: 1916	134
12 The Pathfinder Expedition, Part 3: 1917	140
13 World War I	149
14 The Fight for Naches Pass	153
15 Collaboration	166
16 Movie Making	171
17 The World's Oldest and Youngest Aeronaut	184
18 "No Half Hours to Spare"	190
19 The Oregon Trail Memorial Association	195
20 End of the Trail	212
Afterword	217
Acknowledgments	224
Notes	225
Bibliography	255
Index	260

Map and Illustrations

Ezra Meeker, circa 1926	x
OTMA Oregon Trail Map	xii
La Grande, Oregon, monument in 1906 and 2006	25
Portland, Oregon, in June 1908	28
Meeker and Mardon at Huntington, Oregon	32
Huntington, Oregon, Monument, 2006	32
President Roosevelt viewing the outfit	34
Meeker's 1906 wagon	37
Overview of Meeker's exhibit at AYPE	42
Meeker's restaurant at AYPE	42
Meeker at 1910 International Air Meet	54
Meeker's 1910 camp in Ontario, California	56
Well Spring boulder	63
Baker City, Oregon	65
1911 Columbus, Ohio, dodger	82
Meeker in Texas	88
Meeker and Mardon at the Alamo	91
Meeker caulking his graffiti covered wagon	96
Cover of Meeker's 1912 book	106
Point Defiance glass house	108
Pathfinder in Indianapolis	124
Pathfinder in Seattle	131
Prospectus for Pioneers of America	174
1914 Oakland, California, dodger	178
Meeker's third airplane flight	185
Meeker and President Coolidge	189
Ezra Meeker and Tex Cooper	192
Fort Hall monument plan	196
1926 memorial coin	200
OTMA touring car	202
Ezra Meeker at dedication of his statue	203
Meeker selling coins in 1926	205
Meeker with his 1928 Oxmobile	213
Meeker's 1923 wagon	215
Meeker's grave	216

Foreword
Will Bagley

Ezra Meeker bestrides the legacy of the Oregon Trail like a colossus. He wrote a shelf of books about its history and became the chief promoter of its tale. He first came west on the overland trail in 1852 and in 1906 reversed course and headed east, taking his ox team and wagon from Washington State to New York City and the White House. When he set out, Meeker was once again broke. The elders of a Portland church declined to let Meeker give a fundraising lecture in their chapel, refusing to do anything to "encourage that old man to go out on the Plains to die." Little did the Unitarians suspect Ezra Meeker would spend the next twenty-one years preserving, promoting, and protecting the Oregon Trail.

Meeker could be inspirational, charismatic, and beloved as well as arrogant, aggravating, cantankerous, and obnoxious. When he committed to a cause, whether it was making a fortune or passing a bill in Congress, he could be irrational but unstoppable. He achieved the central theme of the last three decades of his life—the preservation of the national memory of the overland wagon road as the story of the Oregon Trail. Meeker's legacy helps explain why twentieth-century Americans remembered the Oregon Trail and forgot the more heavily used wagon road to California, which had its promoters but none who matched the skill and persistence of Ezra Meeker. At age ninety-seven, Meeker set out on what the *New York Times* calculated would be his "Sixth Transcontinental Journey." The *Times* failed to count his innumerable trips to New York by railroad and his 1924 airplane crossing.

Dennis Larsen's epic biography suggests his subject lived at least four distinct lives over his ninety-eight years. Meeker himself described crossing the Oregon Trail and becoming one of Washington Territory's earliest pioneers. *Hop King: Ezra Meeker's Boom Years*, explains how this human dynamo, who always identified as a farmer, became a Puyallup community builder, agricultural tycoon, and world traveler before hop lice and the Panic of 1893 demolished his fortune. *Slick as a Mitten: Ezra Meeker's Klondike Enterprise* tells how he responded to these disasters. Larsen's *The Missing Chapters: The Untold Story of Ezra Meeker's Old Oregon Trail Monument Expedition* covers the venture that made Meeker famous. Now his concluding volume, *Saving the Oregon Trail*, celebrates Meeker's "grand quest to commemorate the Oregon Trail" from 1901 to his death in 1928.

Larsen delivers meticulous, high-grade scholarship: he apparently examined every one of the 50,000 or so pages of Meeker's papers. He is an effective, entertaining writer. His chapter introductions are models of how to tell a reader what you are going to tell them and he ends each chapter with a cliffhanger or engaging teaser. He is organized from top to bottom, from the structure of the book to the sequencing of the sentences.

Ezra Meeker's ethical humanity discredits one of the worst excuses for past prejudices—the argument that bigotry was the societal standard of the time, so everybody was doing it. Not everybody: his ethical core condemned prejudice. Meeker recalled his friend George Bush had "left Missouri because of the virulent prejudice against his race." After the vast migration of 1852 created near-famine prices for food in Washington Territory, Bush "divided out nearly his whole crop to new settlers who come with or without money—'pay me in kind next year,' he would say." Bush gave Meeker milk for his infant son, yet "George Bush was an outlaw but not a criminal; he was a true American and yet was without a country; he owned allegiance to the flag and yet the flag would not own him." Bush obeyed the law but "the law would not protect him; he could not hold landed property; his oaths would not be taken in the courts of law—in a word, an outlaw and yet not a criminal," all because "he had some Negro blood in his veins." In 1856 Meeker was one of two jurors who voted to acquit Leschi, leading to a hung jury. After a second trial the government legally lynched the Nisqually leader in 1858. What did Old Ezra do about it? In 1905 he published a book, *The Tragedy of Leschi*.

Why should Americans remember the Oregon Trail and the life of its most renowned advocate and defender? First and foremost, they are both great subjects and stories, and who can resist a great story well told? During his long and adventurous life, Meeker had been a printer's devil and Henry Ward Beecher's paperboy, a surveyor, an overland emigrant, a longshoreman, lumberjack, Swamp Place farmer, grocer, miner, railroad promoter, women's suffrage advocate, Hop King of the World, president of the Washington State Historical Society, world traveler, and accomplished writer. In the end, *Saving the Oregon Trails* shows how much Ezra Meeker's life reflected our nation's history and transformations. Dennis Larsen's gifts as a storyteller make it a joy to read.

Ezra Meeker, 1926, location unknown. *Author's collection*

Preface

This is the fourth and final volume of a multi-decade effort to complete Ezra Meeker's biography. Meeker lived ninety-eight highly productive years, making the task of telling his story somewhat challenging. It was almost as if he lived four different lives. We first meet a youthful Meeker who ventured to Iowa in the autumn of 1851, traveled to the Pacific Coast on the Oregon Trail in 1852, and pioneered in Washington Territory through the remainder of the 1850s. He told this story in his various books, beginning with *Pioneer Reminiscences* published in 1905. Next came his middle years (1862–96), the hop years, when growing and brokering this crop made him one of the wealthiest citizens in Washington Territory and built an economic dynamo that helped transform the territory into a state. This story is told in my book *Hop King: Ezra Meeker's Boom Years*. An interlude followed the loss of his hop empire and with it his lifelong identity as a farmer. He briefly experimented with a new career in mining, joined the Klondike gold rush in 1898, and spent four years in the Yukon as the proprietor of a general store called the Log Cabin Grocery. This tale can be found in my book *Slick As a Mitten: Ezra Meeker's Klondike Enterprise*. The years from age seventy until his death, his golden years so to speak, were spent in a grand quest to commemorate the Oregon Trail.

Meeker exploded onto the national stage in 1906–8 when, at age seventy-five, he took an ox team and covered wagon east over the Oregon Trail, setting markers along its path, and continued on to Washington, DC, to meet with President Theodore Roosevelt. I have described this adventure in *The Missing Chapters: Ezra Meeker's Old Oregon Trail Monument Expedition*. This new volume expands on that work, describing Meeker's life beginning with his return from the Yukon in 1901 to his death in 1928, a span that included the monument expedition. It is an attempt to define the man and explain how this unique character captivated the nation around what seemed at the time a quirky idea—saving the Oregon Trail. It is the story of one man's dream, and how that dream became Ezra Meeker's gift to America.

—*Dennis Larsen*

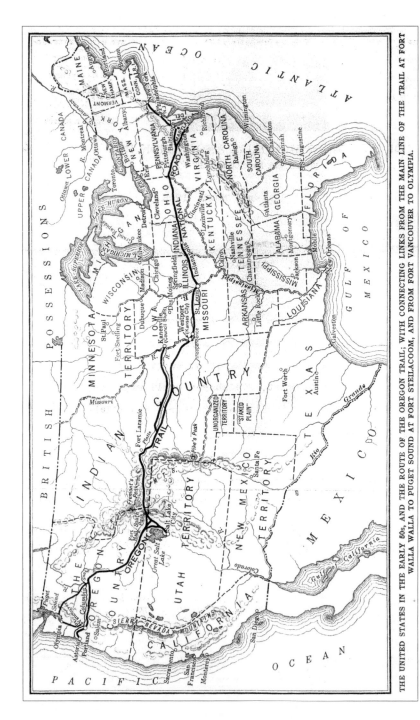

Oregon Trail Map, Oregon Trail Memorial Association. *Author's collection*

Introduction

Ezra Meeker's 1852 trek over the Oregon Trail was an experience burned into his psyche, one he never forgot. By the turn of the century, 1900, the trail was fast disappearing to the plow and development, and the memory of pioneer courage, suffering, and accomplishment was fading from the collective American mind. Meeker was entering his eighth decade and, as one of the few surviving pioneers, he decided to revive that memory.

He had come west with his family as a young man and at first struggled to gain a toehold as a farmer on the shores of Puget Sound. In the final decades of the 1800s he had turned an experimental hop crop into an economic giant, made a fortune, walked with the rich and famous, but was ultimately humbled by financial mistakes and the Panic of 1893, losing it all. However, the essential Meeker remained, the good and the bad. He was blessed with remarkable health and stamina. Personal comfort and the trappings of wealth were not high on the list of things that were important to him. He could live as easily in a log cabin or a covered wagon as in a mansion. His capacity for work was boundless. His drive was unrelenting. He had a strict moral compass and a vision of the future that was unerringly accurate, and he was always willing to embrace the new. But he did not suffer fools gladly, and a pugnacious side to his personality that tended toward self-righteousness grated on many. He was a ruthless competitor. Ezra Meeker was a complicated man.

When Meeker embarked on the project that would eventually define him, he was alone. It was a one-man show. Many thought he was crazy. When his daughter Carrie learned that her seventy-five-year-old father planned to take a covered wagon and ox team east over the old Oregon Trail she was appalled. Knowing his capacity for stubbornness she made an appeal to his vanity, telling him people would laugh at him and call him a foolish old man. It was to no avail. When he approached a Portland, Oregon, minister requesting financial aid, the congregation instead received a sermon from the pulpit urging them to do nothing to aid an old man's wish to go out onto the Great Plains to die. Meeker had no intention of dying. There was work to do. Gradually he pulled those around him into his project, at first just family and close friends. The transition from bystander to supporter was not always a smooth one; toes were stepped on, feelings were hurt. Meeker's intensity and uncompromising determination ruffled a lot of feathers, but over time the circle widened to eventually include governors, congressmen, senators, and even presidents of the United States. They were simply unable to resist the worth of the effort and the unrelenting force of the man.

In many ways this was a unique period in an already event-filled life. Meeker, no longer a farmer or businessman, the roles that defined the majority of his life to this point, transitioned to a new persona, that of the "old pioneer," entertainer and educator extraordinaire, national celebrity, and prolific author. He crusaded tirelessly to remind a fickle and short-memoried public that a good portion of the nation's greatness was built on the backs of the pioneers who ventured westward to anchor the Pacific Coast to the remainder of the nation. Meeker was immensely proud of this accomplishment. He devoted the final years of his long and amazing life to the goals of involving the public, the states, and the United States government in remembering and saving the Oregon Trail. To a great extent he succeeded.

The journey took nearly three decades. He started by writing a book describing his original 1852 trek and his early days as a pioneer in Washington Territory. He made multiple trips over the trail to map and mark it, but in essence providing a living history lesson in the flesh to a nation now experimenting with flight. Meeker traveled the trail three times by ox-team (in 1852, 1906, and 1910) and again in 1916 in a modified automobile with a canvas hood that was probably the nation's first motorized RV. In 1924 he flew over the trail in an army biplane from Washington State to Washington, DC, ending in a meeting with President Calvin Coolidge. He again crossed the continent and the Oregon Trail in 1922, 1923, and 1926 via automobile and railroad. He lectured to tens of thousands, experimented with motion pictures (the new technology of the day) to tell the story of the trail, and sold thousands of books and postcards. He founded a national organization dedicated to memorializing the trail, and helped convince a stingy congress to mint a fifty-cent coin that his trail organization could then sell for a dollar, the profits used to mark the trail. He was poised for another trip in 1928 in the Henry Ford-built "Oxmobile" when he took ill and died. It was the last act of a long history with the Oregon Trail that began in 1852 when this young man who just wanted to be a farmer started west.

Professor of Education in English Howard Driggs,[1] co-author with Meeker of *Ox-Team Days on the Oregon Trail*, in whom Meeker found a kindred spirit, eulogized him at a celebration of what would have been Meeker's one hundredth birthday.

> The great work to which Ezra Meeker clung to the last moment of his life was something far more vital than the mere marking of a pioneer trail. His goal was the preservation of our historical heritage. He strove to keep alive the spirit of real Americanism as it was exemplified in the lives of the heroic men and women who carried America westward. These pioneers knew no such word as *fail*. Ezra Meeker was one of this breed of stalwart Americans. Realizing this, we do not wonder he should lament in passing that his work was not done. One fine thing he did accomplish, however, and that was to plant the cause he so dearly loved in the warm heart of America. It is a cause, indeed, that belongs to all America. In the making of our great West, America was made.[2]

CHAPTER 1

Telling History His Way

Home

Ezra Meeker left the Klondike for his home in Puyallup, Washington, in April 1901, a most dangerous time to travel. This was the time of the spring breakup, and traveling over the Yukon River ice was extremely hazardous. The risk of disappearing forever into the frigid depths as the ice broke up was high. Overland travel through the softening snow was nearly as treacherous. Draft animals often mired in the muck, forcing their passengers to continue on foot at the mercy of spring blizzards. He should have waited until June when the river opened to steamboat navigation. It would have been the prudent thing to do, especially for a man of his age. By the standards of the day, Ezra Meeker, now in his seventieth year, was after all an old man. This was a reality he chose to ignore.

Meeker's Yukon years marked a major transition in his life. He arrived on the shores of Puget Sound in the future state of Washington with his wife and infant son in 1853, nearly penniless. Two decades later he was a wealthy man. By 1896 that fortune was gone, and two years later, in 1898 at age sixty-seven, he tackled the famed Chilkoot Pass and floated down the Yukon River to Dawson City. He made two trips north from Puyallup that year, wintered in the Yukon in a crude cabin, and established the Log Cabin Grocery, a Dawson City fixture for the next six years. He brought his son Fred and daughter Olive north to Dawson along with their spouses. He buried Fred in the Dawson City cemetery. After four years in the north, his son Marion took over management of the store. Another transition awaited.

For public consumption he told everyone he was making this dangerous trip to join his wife for the celebration of their golden wedding anniversary. But he had informed Eliza Jane previously, by letter, that he could not do so. What changed his mind and caused him to undertake such a dangerous journey? His oldest child, Marion Meeker, had arrived unexpectedly in Dawson City, making a dangerous trip of his own north from Puyallup and leaving his family behind. Why, within days of Marion's arrival, did Meeker put his son in charge of the grocery business and start for home?

In the wake of the loss of his financial empire Meeker left an astounding number of debts with a large number of people. The banks had foreclosed on his hop properties but there were many others who were left unsatisfied. Creditors were after his one remaining asset, his home, and likely his son brought news of this to his father. The evidence, while circumstantial, strongly suggests that Meeker returned from the Klondike in April 1901 primarily to deal with this threat, and that being present for his golden wedding was a secondary consideration for leaving the Yukon during the spring breakup.

His arrival home received attention from western newspapers, including the *San Francisco Call*:

> Ezra Meeker…has returned from Dawson to celebrate his golden wedding with Mrs. Meeker at their Puyallup home. At their silver wedding he was a rich hop grower and subsequently became a millionaire owning and operating great hop yards about Puyallup, Sumner, Auburn and Kent. Six years ago hops dropped in price and he gradually lost his farms and business. After the Klondike was discovered he began shipping fruits and vegetables to Dawson. This gave him a start and judicious…investments have again made him financially independent.[1]

He arrived in time to celebrate that fiftieth anniversary with his wife and friends. Soon after his return the Meekers sold their Puyallup mansion to their middle daughter Carrie Osborne and her husband Eben, thus putting it out of the reach of creditors. The three-story edifice was the pride of Puyallup, and was referred to as the "Meeker Mansion." Built at the height of the family's prosperity, it is still a city landmark today. The listed price was $10,000 in gold coin. The payment was to be in fifty-dollar-a-month installments, with the proviso that the elder Meekers retained the option to live there until their deaths if they wished. The Osbornes were to be responsible for taxes and maintenance. Future events, however, clearly demonstrated that Meeker retained control of the mansion property even after the sale.

Upon Meeker's return to Puyallup, his children urged him to stay home and settle down to a quiet life. To his credit Ezra actually tried. He spent the remainder of 1901 and all of 1902 living in the Puyallup mansion. However, putting food on the table required money. Through 1902 Meeker was still sending vegetables north to Dawson City where his son Marion and son-in-law Roderick McDonald were managing his Klondike grocery business. Meeker no doubt received some income from these shipments, but gold fever in the north was dying and with it the vegetable business. Marion came home at the end of 1902 and Meeker's Log Cabin Grocery in Dawson City shut its doors for good. Its demise created a need for a new income stream.

In answer to this need the family began an entirely new business. The mansion's horse stable was turned into a large chicken coop and Ezra and his son Marion started raising Leghorns, selling both the chickens and their eggs locally.[2] There were also forty mature holly trees growing on the grounds and a good business was done each Christmas season selling cuttings from those trees. Supplementing this income were a number of fruit trees and a vegetable garden on the grounds that provided foodstuffs for canning. It was nowhere near the income the Meekers earned during the hop growing years, but it was enough to be comfortable.

Relatives and friends visited often. Both Eliza Jane and Ezra made extended trips to visit their daughters and families in Seattle and Oregon. Youngest daughter Olive and her young son Wilfred came home from the Yukon in September 1902 and lived in the mansion until May 1904. Her husband Roderick McDonald joined them in February 1903 and stayed through the summer before returning to his work in Dawson City.

In the spring of 1903 Meeker wrote to his eldest daughter Ella who lived in Oregon at the time, "I have leased the chickens, house and grounds to Marion but retain the use of the house so we can live here any part of the time we want."[3] Apparently he neglected to remember that in selling the mansion to his middle daughter Carrie in 1901 he had no legal authority to issue leases. This refusal to acknowledge obstacles, and at times reality, in pursuit of his goals was simply a continuation of a lifelong habit. It was a character trait that often got him in trouble. But his pertinacious behavior, while incredibly annoying to many, also made possible accomplishments that seemed, in the beginning, unreachable. And as to his children, they retained the habit of acceding to their father's wishes long into adulthood. Now with two of his children and their families back home, Ezra had the pleasure, for a brief time, of watching his grandchildren grow up. Life was simple and comfortable. It did not remain that way for long.

The Washington State Historical Society

Meeker stated in several letters written to friends during these years that his career as a serious businessman was over. He obviously considered raising chickens a sideshow. It took up little of his time or energy. He was restless and needed to be productive. In 1903 Meeker turned his attention and energies toward transforming the Washington State Historical Society (WSHS) into a viable and lasting institution. He was elected president of the society on February 21, 1903.

Meeker was acutely aware that his generation of pioneers was dying off. With their deaths vanished the collective memory of the many events of pioneer times. Accordingly, he started interviewing and collecting stories and documents from the pioneers and their descendants around the state. He asked the Northern Pacific Railroad to grant him free statewide passage on their trains to carry out this duty. After being rebuffed, he gently reminded them that it was a privilege the company had given him for most of the past twenty years and that he was asking for nothing new. They relented.

Meeker lobbied state legislators through an extensive campaign of letter writing and personal contact, and he secured an appropriation of $5,000 that allowed him to travel around the state using his railroad pass to collect documents and artifacts from the pioneers. He also lobbied for a charter making the society a quasi-state institution. He stayed in Olympia while the bill was under consideration and successfully guided it through both chambers. The governor signed the charter but vetoed the appropriation portion of the bill. Thus the Washington State Historical Society became a state institution operating under a state charter but without funding from the state. The society had a very small office space at this time and nowhere to store a growing collection of documents. Meeker solved this problem by convincing the city of Tacoma to supply a much larger office and a place for the storage of documents in a city-owned building. He also obtained a long-term lease from the city fathers for a nominal rent. The skills Meeker perfected in dealing with politicians in Olympia and Tacoma would, in a few short years, be carried to the national stage, where they were unleashed on the United States Congress in his grand quest to save the Oregon Trail. But for the present, he needed to put the historical society on a sound financial foundation.

Meeker was dogged in his pursuit to replace the funding that the governor vetoed, and he secured a promise of financial aid from the Tacoma chapter of the Daughters of the American Revolution. He also tackled the issue of membership. Oregon's historical society had over six hundred members. Washington's had but forty, and $600 worth of debt.[4] Within a month Meeker had added eleven lifetime members to the society at twenty-five dollars each bringing in $275 in revenue. He was adamant, however, that none of this new money he was raising should go to retiring debt—that it should be used only for the purposes outlined in the bill passed by the state legislature.

The issue of retiring the society's debt created friction between the new president and Secretary Edward N. Fuller and the Board of Curators. (Fuller was also secretary of the Tacoma Chamber of Commerce.) Fuller had collected a large number of obituaries during his time as secretary of the historical society. The Board of Curators pledged to purchase his collection for $600, thus creating debt on the WSHS financial ledgers. Meeker felt this was an improper use of the society's funds. He even hinted at fraud in one of his letters, and he argued that those involved in accumulating the debt had a personal responsibility to pay it back. This advice was not well received.

Ultimately Meeker's vision for the society's fundamental mission was at odds with that of Secretary Fuller and several of the members of the Board of Curators. Meeker objected to the fact that Fuller was still spending most of his time collecting obituaries. He felt the focus should be on obtaining documents and manuscripts from the fast disappearing pioneers. He took his concerns to Arthur B. Warner, superintendent of Tacoma schools and an influential member of the historical society, with disastrous results. Meeker wrote to Molly Male (an in-law and Tacoma school teacher) how the meeting went. He described the atmosphere that greeted him as "like a hoar frost." Warner was of the opinion that preserving the early history of the Indian war, for instance, had no value because it was a scene that had been enacted time and again all over the continent and was of no interest to any except the participants, an interesting perspective for a historical society. Meeker vehemently disagreed. He told Male, "I came away with indignation boiling in my veins."[5]

The board also ignored him on the question of finances and voted to use the money he was raising to retire the debt. Meeker responded by informing Secretary Fuller that he would no longer solicit new members.[6] That vote was the breaking point for Meeker. On May 11, 1903 he wrote Allen Weir (Washington's first Secretary of State and president of the Thurston County Pioneer Association), "Things have gone wrong in our society of which I wish to tell you when we meet…I have ceased to make farther effort for the society."

Throughout 1903 Meeker found time for other activities. Washington was marking the fiftieth anniversary of its becoming a territory. The focus of the celebration was in Olympia. Meeker was selected to speak on behalf of the pioneers at the Olympia Opera House. His topic was "Retrospect." On April 2, 1903, he addressed some two hundred Washington teachers at the annual teachers institute despite the objections of Superintendent Warner. On May 23–24, 1903, President Theodore Roosevelt made a visit to Puget Sound. A highlight of that visit was the presidential boat excursion from Tacoma to Seattle. Meeker was invited on the excursion as the representative of the "Old Settlers." Meeker felt there should have been others of his generation with him on the boat but he lost that fight. This was Meeker's first meeting with President Roosevelt. Four years later he would renew the acquaintance when he drove his covered wagon and ox team down Pennsylvania Avenue to the White House.

Pioneer Reminiscences and the Tragedy of Leschi

The project that would consume the next two years of Meeker's life and that would become part of his lasting legacy to the state of Washington was started on the very day he became the president of the Washington State Historical Society. He described it to his daughter Ella. "I am actually at work on a book to be entitled 'Fifty years of Pioneer life,' a history of early days here and am devoting my whole time to it. It is a big job for me…and if I get through in a year I will think I will do well."[7]

By May 4 Meeker had fifty typewritten pages ready to be proofread. In typical fashion he was thorough. The first months of the work were devoted to research. He wrote dozens of letters to living pioneers who were actively involved in much of the history that was his subject. Eliza McAuley Egbert, Nelson Sargent, Edward Huggins, Lulu Packard, and his brother John Valentine Meeker (living in the Oddfellows home in Walla Walla at this time), were among his early correspondents. Meeker's favorite research correspondent seemed to be Edward Huggins, a former Hudson's Bay Company employee whom he pumped about activities of the Puget Sound Agricultural Company and the Hudson's Bay Company. He also started gathering government documents from Olympia to Washington, DC.

On May 19, 1903, Meeker listed the Puyallup mansion and grounds for sale with the Seattle firm of Crawford & Hanover, again ignoring the fact that it legally belonged to his daughter. Ezra and Eliza Jane now began a routine that would become commonplace during the next two years. They spent long periods of time living in Seattle in the Osborne household. From there, Meeker made extended excursions to Portland to study records held by the Oregon Historical Society and to Bellingham, Port Townsend, Tenino, Centralia, Olympia, and elsewhere, to personally interview as many of the old timers still living as possible. Some of these interviews lasted two or more days. A few pioneers were brought to Seattle at Meeker's expense and put up at the Osborne home.

When the expenses of doing the research began to tax his modest financial resources and the effort to sell the mansion failed to bring forth a buyer, Meeker told Robert Wilson (his accountant, tax advisor, and real estate agent) that he was open to renting the mansion, but not for a "low price."[8] He also considered moving to Seattle permanently and building a house next to his daughter, but this did not happen.

He renamed "Fifty Years of Pioneer Life" a short time later, calling the book the "Tragedy of Leschi." Chief of the Nisqually tribe, Leschi was tried and executed for murder following the 1856 Indian wars. Meeker always argued that Leschi was wrongfully convicted and that he was "judicially murdered." Much effort was expended in creating an illustration of Chief Leschi since no contemporary portrait existed. Meeker sent a description, apparently from memory, to the artist who was to sketch a portrait of the Indian chief for the book.[9] After obtaining the sketch Meeker sent a copy to Edward Huggins and others, including Leschi's wife, asking them to recommend needed changes.

He became so absorbed in his work that he could barely take time away for family affairs. Meeker was in Seattle and Eliza Jane was in Puyallup when the following was penned, "Now Mamma, I love you. I do, but I cannot stop my work to move furniture however."[10]

By early November 1903 Meeker was anticipating a completion date between January and March 1904. Even at this point, with much of the book written, he was still reaching out far and wide for any historical information he could get, but he now faced the hurdle that all writers face. Who would publish the work? His hope was to sell the book by subscription on the West Coast and find a reputable East Coast publisher to also print a version. He priced the book at three dollars and felt he would need to raise around $2,000 to justify printing.[11] Pledges to purchase the book, or subscriptions, would allow him to present evidence to a publishing firm that if they took on the cost of printing, no financial loss would accrue. Meeker felt he would need about one thousand subscriptions to interest a publisher in the project. He wrote to family members and friends around the state asking if they would be interested in soliciting subscribers in their hometowns for a commission. He also sent prospectuses to the editors of several northwest newspapers such as the *Tacoma Ledger* and the *Oregonian,* hoping for some free publicity. He even prepared a small booklet to be used by his canvassers or agents to advertise the book to prospective buyers. He vowed "I will not put out a work that is not first class as to printing & binding."[12]

By mid-February 1904 subscribers had pledged a total of $500, and Meeker concluded that it would take another year to secure enough advance subscriptions to convince a publisher to print the book. An impatient Meeker wanted to move at a faster pace. He decided to self-publish. He wrote his accountant Robert Wilson asking for a short-term loan to cover the cost of publishing.[13] Wilson declined.

Meeker was also quite aware that the book would be controversial, especially for his depiction of Governor Stevens. "I am admonished by some intimate friends that I may expect my work to be 'jumped on,' as they express it, by the friends of the former Governor. I tell them I have simply written the truth and if it hurts anybody, I can't help it."[14]

Meeker also kept another potentially explosive issue confined to his correspondence at first. Since it was Ezra's belief that the Medicine Creek Council and subsequent treaty triggered the Indian wars, he wanted to learn all he could about what went on there. Even today the sequence of events, and who said and did what, are controversial.[15] Meeker interviewed everyone he could reach who had lived through those times, but when he went looking for the official government documents that were stored in Washington, DC, he learned they were missing. After much searching he discovered that Hazard Stevens was the last person known to have consulted them when he wrote a glowing book about his father. Meeker suggested that perhaps what was in those documents was so damning to Stevens' reputation that his son made sure they simply disappeared.[16] Meeker's portrayal of Hazard's father as an alcoholic who was drunk much of the time during the crucial events of 1855–58 would indeed ignite fireworks.

The effort to obtain financing for the printing of the book stalled and he told his correspondents that he now hoped to publish in June 1904. There is a long break in the Meeker letters from March to November 1904. The first November letters are about an effort on Meeker's part to bring a fruit cannery to Puyallup and the chicken business that he had turned over to his son. Not until November 19, 1904, is there mention of the book, and in these letters Meeker is trying to unravel the story of the disappearance of the Medicine Creek documents. There is no mention of publishing.

Problems at the Mansion

By late 1904 the Meekers had moved out of their Puyallup home temporarily and were living in the Osborne household in Seattle while Ezra rented out the mansion to boarders to help meet expenses. Carrie and Eben Osborne exercised little or no say in the management of the property at this time. One assumes they agreed to the transfer of ownership simply to eliminate the threat of Meeker's creditors, without expecting a transfer of control. Meeker acted as their de facto manager, but with his own best interests at heart.

The renters were four families aggregating fourteen people, paying forty-five dollars a month total in rent, and Meeker found the business of being an absentee landlord difficult. He had placed Mrs. Eva Gear in charge of managing the mansion in his absence, but the arrangement did not work well. He dismissed Gear in December,[17] and beginning in January 1905 traveled weekly from Seattle to Puyallup to spend each Monday and Tuesday managing the affairs of the mansion and dealing with the tenants. Katherine Graham replaced Gear and remained in that capacity until late 1906 or early 1907.[18] Meeker also increased the number of holly trees to over a hundred and had a hundred chickens wandering the yard and scratching under those holly trees. His nights in the mansion were spent sleeping in the parlor or attic.[19]

PUBLICATION

As 1905 dawned Meeker was nearing the completion of his book. He planned to sell copies at the upcoming Lewis and Clark Exposition in Portland, Oregon, but had not yet secured financing for the actual printing. Louise Ackerson,[20] an old family friend, came to his rescue. Her husband John Ackerson, in partnership with others, had built a sawmill in Tacoma in 1868 that made the couple quite wealthy. Louise and her husband purchased property in Puyallup in the 1870s and were neighbors to the Meekers for several years. When the Meekers went to the Chicago World's Fair in 1893, Louise, now a widow, went along. In late 1904 Louise decided to sell forty acres of her Puyallup property and asked Ezra to act as her real estate agent. The negotiations were complicated. Several business letters went back and forth between Meeker and Mrs. Ackerson who was then living in Los Angeles and later in Honolulu. At the end of December, after successfully selling her Puyallup property, Meeker wrote a long letter describing his efforts to sell subscriptions. He concluded by asking her if she would consider financing the printing of the book.[21]

Mrs. Ackerson expressed an interest in helping if he would include some items that she desired to see in print. Meeker immediately agreed. Thus came the chapters titled "Pioneer Religious Experiences" and "The Morning School."[22] With her wishes fulfilled Mrs. Ackerson offered to cover half the cost of the printing and suggested that Mr. Osborne pick up the cost of the other half. Meeker sadly informed her that Mr. Osborne was unable to make such a loan. He told her "I may be able to get the remainder elsewhere but do not know just where; I can only say I will try."[23]

Help came from an unexpected quarter. The Metropolitan Press Company of Seattle agreed to take on the other half of the loan.[24] Meeker's long-time attorney John Hartman Jr. drew up the contract. The copyright, as collateral for the loan, was assigned to Eben Osborne in trust for the two signatories. However, the actual printing of the book was done by the Lowman & Hanford Company. Meeker leaves us in the dark as to how the second loan was secured and why the Metropolitan Press allowed another company to do the actual printing. Perhaps the owner of the press recognized that the firestorm of criticism ignited by Meeker's depiction of Governor Stevens as a drunk might prove to be good for sales but an uncomfortable endorsement for the company.

In a February progress report to Mrs. Ackerson, Ezra reported that Hazard Stevens acknowledged looking at the missing government documents but claimed the clerk he worked with must have lost them. Also, space considerations forced Meeker to edit down events that took place after 1859, although he included a chapter on the origins of local place names, giving Louise Ackerson's husband credit for naming Tacoma. "I hear that Hanford, who has been reading the proofs is highly pleased with the work," Meeker wrote, "[I] will send you the first copy of the book I send out."[25]

Lowman & Hanford made a rather unique arrangement with Meeker in regards to the book now titled *Pioneer Reminiscences and the Tragedy of Leschi*. They printed two thousand copies, but bound only two hundred of them. A few of the bound copies were put on sale at their store in Seattle, and the remainder were given to Meeker to sell. When he was about to exhaust his supply, Meeker would request that another two hundred copies be bound and shipped to the addresses he supplied. The binding fee was thirty-five cents a copy. The boxes of unbound books were stored in the company warehouse.

Although the two loans were paid off in 1907, it took years to sell out the edition. Meeker made an effort in the summer of 1905 at the Lewis & Clark Exposition in Portland. He found sales slow. He continued to sell the book along with his newest and more popular work, *The Ox Team or the Old Oregon Trail*, throughout the 1906–08 Old Oregon Trail Monument Expedition. Sales went on through the Alaska-Yukon-Pacific Exposition in Seattle in 1909 and during his 1909–10 southern California expedition. Meeker was still at it on his second Oregon Trail Monument Expedition from 1910-12, selling from his covered wagon. On October 11, 1911, there were still eight hundred unbound copies stored in Lowman & Hanford's warehouse. Meeker had copies of the book for sale as late as March 8, 1919, when he arranged for J. K. Gill and Company to sell the book in their stores.

Pioneer Reminiscences of Puget Sound: The Tragedy of Leschi is today considered a northwest historical classic. The original publication can be found occasionally in bookstores that specialize in rare used books. It has gone through other printings since Meeker published his hardcover copy in 1905. The latest paperback reincarnation split Meeker's original volume into two parts, one titled *Pioneer Reminiscences* and the other *The Tragedy of Leschi*.[26]

CHAPTER 2

A War of Words, a Test of Wills

MEANY VS. MEEKER

Meeker's book was in print by early April 1905 and, as predicted, it immediately ignited a firestorm. Edmund Meany, head of the history department at the University of Washington, declared war on Meeker's work in the pages of the *Seattle Times*. Headlines such as "Pioneer Has Failed to Tell the Truth," "Advances Falsehood," and "Charge of Malice Too Mild," blared out from the pages of the *Times* for days. Meeker did not remain silent in the face of withering criticism. He fired back. Meany started the war of words stating,

> Everyone will praise the first part of Mr. Meeker's book—his Pioneer Reminiscences—and the concluding chapter by Clarence B. Bagley shows a vast amount of careful research among old records. Between these two meritorious portions is the 'Tragedy,' which deserves criticism because it assails in ruthless fashion the character of a man whose memory has been cherished in love for nearly half a century.[1]

Meany continued in the same issue of the *Times* to attack Meeker's charge that Leschi did not sign the Medicine Creek Treaty. "He has for years been giving currency to a lot of Indian stories relative to it, with the hope, apparently that they will not be denied and at length be considered true history." The professor claimed nineteen white witnesses testified that Leschi signed the treaty and all that Meeker had was a collection of "statements from a number of Indians to the effect that Leschi did not sign the treaty." Meany's assumption, of course, was that the testimony of the white witnesses trumped that of the Native Americans. Meany also attacked Meeker's argument that the smallness and poorness of the reservation was the cause of the trouble and he railed against Meeker's portrayal of Stevens as a drunkard calling it "the most nauseating part of Mr. Meeker's book." The professor argued, "It would be impossible for a man to be a diseased inebriate and to work out the wonderful career of achievement that stands to the credit of Isaac I. Stevens.... Any freshman in college who would seek to bolster up such a grave charge with such flimsy and reprehensible statements would be marked zero on his composition."

This was not the first time Meany and Meeker had clashed. In 1891 Meeker was appointed the Commissioner of the Washington State exhibit at the Chicago World's Fair. Meany was on the Board of Directors. The two men fought over the control of finances. Meeker went so far as to accuse Meany of financial impropriety. That dispute was also fought on the pages of the Northwest press. The end result was that the board, under Meany's leadership, removed Meeker from his position as commissioner. No doubt remembering that event well, Meeker quickly responded to Meany's attack.[2]

> In the *Times* of April 15 Prof. Meany of our state university calls me a liar.... The professor also says I am mean. I published the well-authenticated fact that Gov. Stevens was an inebriate. Prof. Meany says that I lie, forgetting that the use of opprobrious language is not an argument.
>
> If Prof. Meany wants to know the truth, I can take him to a gray-beard living in this city who dined in Olympia during the Stevens administration, who will tell you he repeatedly saw Stevens "beastly drunk," but does not want his name made public. Before going to this man I would want one other stipulation and that is that Prof. Meany will not call the gray-beard a liar to his face, for some old men may forget they are not still young and I do not want to be the witness of a scene.[3]

Meeker offered in the *Seattle Times* article to take Meany to Olympia where five more witnesses would support the charge that governor Stevens was "a confirmed inebriate" and that George Gibbs told the truth fifty years earlier when he wrote that "Stevens, in his fits of intoxication knew no bounds to his language and his actions."[4] In response to Meany's arguing that the Medicine Creek Treaty confining the Nisqually tribe to an unsuitable reservation did not cause the trouble, Meeker pointed to other tribes such as the Snohomish, Snoqualmie, Chehalis, Lummi, and Clallam that did not go to war, maintaining it was because they were given appropriate reservations. Meeker claimed that fully nine-tenths of the Puget Sound Native Americans refused to go to war and remained friendly through the whole trouble, and that the three tribes that did go to war "had a real grievance in that an attempt was made to rob them of their homes without giving a place where they could even build a house, so dense was the timber." He went on to say he lived just a few miles from the council grounds and knew the local Indians well "and no one need tell me that the Medicine Creek Treaty was not the sole cause of the war, for I know better." He concluded that Governor Stevens was responsible for the harsh terms that caused the discontent and that he believed that the governor would not have so acted had it not been for his habit of intoxication.

Meany shot back repeating his argument that it would be impossible for a habitual drunkard to achieve the "great works" that were attributed to Stevens. "I am willing to stake my reputation as a man and as an historian that Isaac I. Stevens…could not have been a diseased inebriate or habitual drunkard."[5]

The Lewis and Clark Exposition

Meeker did not stay around to finish his war of words with Professor Meany. He was bound for the Lewis and Clark Exposition in Portland, Oregon, with a covered wagon and the hope of obtaining an ox team. It would be his first endeavor using this form of travel since 1852, and it had its rough patches. His plan was to set up a pioneer exhibit on the exposition grounds, sell his book, and use the journey to Portland as a dry run for his upcoming Old Oregon Trail Monument Expedition. Meeker had gone public with his plan to take a covered wagon and ox team east over the Oregon Trail, and it was widely reported in various newspapers of the day.[6] Thomas McAuley, a member of Meeker's original 1852 wagon train, living out his days in California, sent Meeker a local newspaper clipping telling of Meeker's grand plan and gently ribbed him, suggesting that Ezra could not handle an ox team in 1852 and what made him think he could do better now.

Meeker's experience that spring and summer would be repeated over and over again during the next decade. Travel by covered wagon, camping outdoors, selling his books and literature, lecturing to local audiences, fighting to get into venues, and his refusal to acknowledge the limitations of his age defined his life from 1905 to 1917. His stubbornness and prickly personality, at times, made things more difficult than they needed to be. But that pugnaciousness also carried him through the low spots in his journey and allowed him, in the end, to be successful beyond his wildest dreams.

On May 27, 1905, accompanied by "helpers" Herman and Tillie Gobel,[7] Meeker started slowly south for Portland with a horse team pulling his wagon since suitable oxen had not yet been found.[8] Arriving in Chehalis on June 1, Meeker bought a pair of oxen named Stub and Twist for $112.50 and spent the next morning working them in the yoke. He shipped the horse team back to Puyallup and his books and advertising matter ahead to Portland. It took another day to properly fit harnesses for the oxen; then Meeker had a photograph of himself and the new outfit taken. Working his way south to Toledo, on June 5 Meeker found a meeting of the State Grange in progress, and camped. Nearly half the town came out to see the outfit. In the evening a reception was given in Meeker's honor and he gave the first of many "pioneer talks" that became a staple in his future travels. Meeker ended his overland travel here. The oxen, wagon, and all hands were put on the steamer *Northwest* and shipped to Portland.

On the morning of June 7 they were unloaded from the steamboat and driven to the exposition grounds. Meeker left the wagon on the street and pitched his tent on an adjacent property where there was grass for the oxen.[9] Then he marched off to see Harvey Scott, editor of the *Oregonian*, and George Himes, secretary of the Oregon Historical Society. He secured letters from each to the president of the exposition asking for Meeker's admission to the grounds with his outfit and the right to sell his book. Initially, he was received favorably by the exposition president and promised the concession he asked for, then told to find a location for his camp.[10]

Roadblocks

Meeker's credentials were impeccable. He had been the commissioner for the Washington Territory exhibit at the 1886 New Orleans exposition and he served in the same capacity for the 1893 Chicago World's Fair until Edmund Meany squeezed him out. He was well known in northwest business circles as the "Hop King" of the United States. And he had a close relationship with Portlanders George Himes and Harvey Scott. But he had been absent from the world of powerful men since his bankruptcy in 1896. From 1896 until 1901 he had even been out of the country. Now Meeker was bringing an exhibit recalling the past to an exposition that promoted the future, and not everyone saw it as a comfortable fit.

The process that started smoothly quickly became mired down in exposition politics. Meeker found a suitable location on the exposition grounds but when he returned to see the president he found his access blocked by some of the lesser officers. The next day the situation repeated itself, so he remained "in camp two blocks from admission gates of exposition, grass, water are plentiful and weather fine."[11] This was good training for an impatient Meeker. In the next few years he would often find roadblocks between him and his goals. He usually found a way to overcome them.

While waiting for the exposition authorities to make a decision Meeker visited eighty-one-year-old Mrs. Lulu Packard at her daughter's Portland home. The two pioneers reminisced about Lulu's arrival in western Washington in 1854 after crossing the plains with Ezra's parents and brothers. Meeker wrote, "The most vivid recollection of an incident was the loss of a Salmon breakfast cooked by herself by upsetting on the sand; she was so hungry for it she could not be reconciled to the loss and added she was not reconciled to the loss yet."[12]

On June 12 Meeker wrote exposition President Henry W. Goode complaining about his treatment. Officials would sneak out back doors leaving a steaming Meeker waiting an hour in an outer office. Eventually he was told he could have a space in "a place way out, where [it] would do me no good." Meeker argued to Goode, "I am certainly entitled to courteous treatment if no more…I am entitled to a decision at your hands and that as I have made the application in writing I wish to know if that decision is your decision and that I am to be ruled out, and if so am I not entitled to know why."[13]

The next two days were spent fruitlessly awaiting a reply. On June 16 a chagrined Meeker met William D. Fenton, an exposition board member, who advised him that he would be admitted to exhibition grounds with his exhibit of pioneer life if he would eliminate the oxen. Meeker conceded.

Success

On June 24 Meeker stated in his journal, "All objections to my occupying the big stump (13 feet in diameter) at the south east corner of the Washington State building removed by leaving my team, tents and wagon on the outside and putting my other pioneer exhibits on the stump." The next three days were spent arranging the exhibit. The oxen were shipped north to Toledo, Washington, and put in the care of Dillon S. Farrell.[14] In late September Meeker shipped them with instructions for their disposition to Katherine Graham.[15] By the end of the month Stub and Twist were eating their hay and alfalfa in Puyallup. And while Meeker abandoned the idea of displaying an ox team at the exposition, he did not give up on the covered wagon. More letters to other officials finally opened the door to allowing the wagon into the exposition grounds.[16]

Meeker set up shop in a hollow stump. No photographs exist, but we can assume it was a quite large hollow stump because he used it as a reception room to display his book and to entertain. A simulated campfire was made by placing a red electric light under a pile of dried droppings from the oxen Stub and Twist. Wrote Meeker, "The Oregon historical society will loan me all the old time relics I can use including a big coffee mill to represent your grandfathers mill he used to grind his meal. I sold two books yesterday but to go out and canvass seems to be an uphill business yet I will doubtless sell many more to visitors of my camp—can but try."[17] A few interesting journal entries are notations made about pioneers who visited with him at his exposition exhibit. Two of particular note follow:

> Mrs. N. M. Bogart of Renton, Wash. crossed plains in 1843. [Marcus] Whitman took off all clothes but underwear and led front yoke of oxen of front wagons across Platte river and saved the wagons from getting into deep water. Crossed the river with a sort of pontoon bridge floated with barrel and boxes & c.[18]
>
> J. P. Eckler; one of the jurors in the second trial of Leschi; voted against convictions at first and two others but finally persuaded to acquiesce. Was at battle of White river and believes I am right in my version of that battle…was all through the war and says my history is correct.[19]

The second entry clears up a chapter in the story of Chief Leschi. His first trial resulted in a hung jury with Meeker and one other juror refusing to convict. This trial was held in Steilacoom. The second trial was moved by some rather suspect means to Olympia where it was assumed there was more favorable ground for a quick guilty verdict.

Thus the length of the deliberation puzzled many spectators when the jury met from 11:00 p.m. until 10:30 the next morning. The note above from Meeker's journal offers an explanation.

On October 14 Meeker shipped his covered wagon and camp outfit to Puyallup. He recorded in his journal, "Left Fair Grounds for good and all after 4 months and 7 days attendance in which time I have not slept under roof or eaten but few meals outside my camp. I take the 11:45 night train for Puyallup where I will arrive in early morning."[20]

The final accounting, after listing expenses of $264 combined with his book sales, showed a net balance of sixteen dollars cleared for his stay at the exposition.[21]

Three months later he would be on the Oregon Trail again, after half a century, with a new wagon and one new ox, beginning a two and one-half year adventure he called "The Old Oregon Trail Monument Expedition." Meeker was about to find his calling.

CHAPTER 3

The Old Oregon Trail Monument Expedition

No part of the Ezra Meeker story has been covered quite as thoroughly as that of the Old Oregon Trail Monument Expedition. Meeker wrote the first of his many versions of the 1906 story as he was traveling east. He began writing in earnest as he neared South Pass. In late July he closeted himself at North Platte, Nebraska, and finished the work with the exception of the final three chapters. By the end of September he was selling *The Ox Team or the Old Oregon Trail 1852–1906* from his wagon. The book eventually went through four printings selling around 19,000 copies. The fourth edition expanded considerably on the previous editions. Meeker continued telling his version of the story of the Monument Expedition through his next three books: *Ventures and Adventures of Ezra Meeker* (1909), *Personal Experiences on the Old Oregon Trail Sixty Years Ago* (1912), and *The Busy Life of Eighty-Five Years of Ezra Meeker* (1915). Each added a few details not present in previous works. In his book *Seventy Years of Progress in Washington* (1921) he added yet more details of that 1906–1908 adventure. *Ox-Team Days on the Old Oregon Trail* (1925), written in collaboration with Howard Driggs, told the story once again, but aimed at a younger audience.

All of Meeker's published works told what he did. None told how he accomplished it. Here we look at a few of the lesser known aspects of Meeker's grand plan and perhaps put into focus what he actually accomplished by driving an ox team from Puyallup, Washington, to Washington, DC.

ORIGINS OF THE MONUMENT EXPEDITION

It is not certain when the idea of the Old Oregon Trail Monument Expedition took root in Meeker's mind. It was surely present before he went to the Lewis and Clark Exposition in 1905, as the story of his plan to cross the continent with an ox team and covered wagon appeared in newspapers across the country that year. In June 1905 he received a letter from Mr. and Mrs. Eben Ives in response to a story about his plan that appeared in an Omaha, Nebraska, newspaper:

> We have just read in the 'Omaha World Herald' an article by, 'W. A. Beindley' that tells of your self and the trip you are planning to begin in July—The crossing of 'The Plains,' by ox team, over the old Oregon route. We are so deeply interested in this scheme, and the success of the object that we cannot resist the desire to write you and wish you 'God Speed,' and may all good fortune attend.[1]

Later in the month Meeker received a letter from a Philadelphia press bureau requesting details of the proposed July 1905 trip along with some photographs.[2] However, the idea of a July 1905 departure quickly disappeared. In mid-August 1905 Meeker wrote Amos Barrett that he now intended to start in the spring of 1906 and stated, "My physical ability to carry out such an enterprise, that is, to personally cross the continent again, is beyond question."[3]

The goals of the expedition were twofold: first to mark the fast disappearing trail with monuments of some kind[4]; second to perpetuate the memory of the pioneers who came over that trail and what they accomplished. Over the course of the journey, Meeker gradually came to embrace a third purpose. This was the building of a transcontinental highway, a national road, along the route of the trail. He even had a name for it—Pioneer Way.

Planning

Meeker's various books give the impression that the Monument Expedition evolved naturally as the covered wagon and ox team headed east—that the townsfolk along the way spontaneously jumped to aid Meeker's cause. And it seemed that all this happened with very little advance planning. The correspondence tells a different tale. He planned meticulously.

Meeker's intention was to spend the first two weeks of February 1906 lecturing in western Washington cities. Speaking venues were obtained in advance, as were the one hundred slides that he used to illustrate the history of the trail in his lectures.[5] He obtained and

La Grande, Oregon, monument in 1906 and 2006. The Oregon Trail went up the hill behind. *Author's collection*

learned to use a stereopticon, an early version of a slide-projector. He planned a monument fund and solicited directors to supervise it. The first $500 raised was to be earmarked for the erection of monuments along the Cowlitz Trail in Washington.[6] He sent letters to prominent citizens living near the selected monument sites arranging for them to take the lead in actually getting the monuments erected. He intended to do the same for Oregon. Later, as he traveled east, Meeker often went ahead to make advance arrangements while his helper followed with the covered wagon and ox-team.

The correspondence between Meeker and Mrs. Roxie Shakleford and Elizabeth Laughlin Lord at The Dalles, George Himes at Portland, and Lee Moorhouse at Pendleton, Oregon, suggests the depth of his planning. No topic was overlooked. He told Mrs. Shakleford,

> I will arrive in The Dalles dressed in the garb of an old pioneer and with an old time prairie schooner wagon, one hub of which did service across the plains in 1853.[7] My talk will be reminiscent in part and historical in part following illustrations on the stereopticon. There will be some old time songs shown on the screen and the audience invited to join, the music to be played on my old melodeon (the first in this country) where a lady can be found to play and every effort will be made to make the meeting pleasant as well as instructive.[8]

He informed Mrs. Shakleford that he had a stove in the wagon for cooking and heating, that the canvass top would protect him from the elements, and that he would have two men with him to help with the camp chores. Meeker even sent photographs ahead to the local newspapers to ensure that his arrival was covered. He also planned to write a book about his journey. He expected to spend the winter in Indianapolis where his wife would join him, and he wanted to be there before the first snow.[9] To George Himes he wrote, "Fortunately I came with some funds of my own which renders me quite independent for three months on the road provided I do not buy a horse team but I want one awful bad but I am keeping out of debt so far and think I'll rather continue."[10]

To Lee Moorhouse he inquired of the weather conditions in the Blue Mountains, wanting to know the snow depth and if it was feasible to force his way through "by shoveling snow and tramping down." Meeker also requested a list of towns between The Dalles and Pendleton where he could get a hearing, and finally, he asked where he could get feed for his oxen.[11] He envisioned that future meeting in Washington, DC, with President Roosevelt and prepared the groundwork. "I venture to enclose prospectus of my work, Pioneer Reminiscences of Puget Sound, etc., which is also a history of the Indian war of 1855–56. Should you be interested in the subject I would feel honored to place a copy in your hands."[12]

Oxen and Helpers

Meeker's oxen, Twist and Stub, were returned to Puyallup in October 1905 from their pasture near Toledo. This is the last we hear of Stub, his fate unknown. Twist's new yokemate for the Monument Expedition was a steer fresh from Montana purchased from the Carston Packing Company in Tacoma for $70.43. The ox was initially named Davis after Henry Clay Davis of Claquato, Washington, who donated fifty dollars of the cost, but he quickly became Dave—a mean, contrary, and somewhat lazy yokemate to Twist. Unfortunately, Twist ate some poisoned feed near Brady, Nebraska, and died on the trail. To replace him, Meeker found Dandy in a local stockyard in Omaha, Nebraska. Dave and Dandy would be yokemates until 1914 when Dandy died after a drive from Seattle to Portland. Dave met his end in California in 1915.

Hiring didn't seem to be a problem, but hanging onto a crew gave Meeker headaches. Camping outdoors in the Northwest in February could be challenging and more than a little unpleasant. It was cold, it rained continually, living conditions were primitive, and none of this inspired loyalty to the mission. Herman Gobel and his "bride" Tillie signed on for a second stint with Meeker, having accompanied him to the Lewis and Clark Exposition the previous year. They had five-month old Juanette Gobel with them. Tillie and the baby bailed out in Portland. Herman lasted until The Dalles. Ed Sorger lasted but a day in Puyallup. He was replaced by William Hayes, who quit at Portland. In Chehalis, Meeker meet a "Mr. Hicks" who joined the expedition at The Dalles and was fired by Meeker in Pendleton.

While Meeker was camped at The Dalles, Oregon, a thirty-two-year-old drifter named William Bruce Mardon stuck his head through the door of the frosty tent and asked for a job. Meeker didn't much like his looks, but desperation is an excellent motivator. He hired Mardon on the spot, and the arrangement stuck. Mardon took over the camp chores, drove the oxen, and helped Meeker sell his books and postcards.

Often Meeker traveled by train to stops down the road where advance arrangements were made. Mardon was left to bring the wagon and ox team forward to meet him. Meeker found Mardon loyal and reliable, but not easy to get along with, and the close living arrangements grated on him. Mardon had a temper and apparently Meeker found some of his habits crude. The forty-three-year age difference between the two men did not help matters. However, it is fair to conclude that without Mardon, the expedition would have failed, and Meeker quickly came to realize that he needed Mardon if he was going to succeed.

Meeker was thrown off stride in 1907, but only for a week, when Mardon announced that he had married Cora Miner of Johnstown, New York, after a courtship of just days.[13] Meeker assumed he was about to lose his indispensable assistant. Instead he acquired a new helper at no extra cost, who also happened to be an excellent cook. Cora and William Mardon stayed with Meeker from September 1907 through early 1910. Health issues plagued Mardon throughout 1910 and he was with Meeker for just parts of the year. In 1911 he, and at times Cora, were with Meeker. Mardon was again with Meeker for part of his 1912 campaign before leaving for good—this time to Stamford, Connecticut, where he lived out the remainder of his life.

Portland, Oregon, in June 1908; Cora Mardon in the wagon with the collie Jim, William Mardon standing by the oxen; Meeker and George Himes standing by the wagon. From Meeker's book *Ventures and Adventures*.

Meeker found a companion in a collie named Jim. He acquired the dog in Puyallup just before he departed that city. Despite having more hair-raising misadventures than were healthy for him (or Meeker), Jim made it all the way across the country and back and even signed up again for a second transcontinental journey in 1910.

ELIZA JANE

There is no correspondence written by Eliza Jane Meeker in the 50,000 or so pages of the Meeker collection, although there are many letters written to her by her husband. It is through these letters that we get our only hints of her thoughts and feelings. The Puyallup Historical Society at the Meeker Mansion has contacted family descendants over the years, but nothing has been found. Unfortunately, by accident or design, Eliza Jane's voice is silent. The only document in the collection in her handwriting is a brief daybook of an 1890 trip to Washington, DC, where she, Ezra, and their daughter Olive attended the National Woman's Suffrage Convention. It tells us that Eliza Jane knew Susan B. Anthony and other national leaders of the woman's suffrage movement, that she attended the business meeting of the national organization, and that she had an appointment at the White House in 1890, the details of which are told in *Hop King: Ezra Meeker's Boom Years*.

Mrs. Meeker learned to live with long separations from her husband. His pattern of taking extended absences from home began almost immediately after the wedding. In a letter to his daughter years later Meeker wrote, "In a few days [after deciding to move to Iowa] your mother went back to her fathers for the harvest soon came on and I had an interest in a threshing machine and was away from home most of the time."[14] Their first winter as a couple found Ezra wandering around Iowa on a surveying crew while Eliza Jane stayed in a one-room rental in Eddyville. In 1853 he left her in their Kalama cabin in northern Oregon Territory for weeks while he and his brother Oliver scouted Puget Sound, looking for a potential home. The next year Meeker left Eliza Jane in their new cabin on McNeil Island while he went alone over Naches Pass to meet his father's wagon train and guide it safely over the Cascade Mountains. In 1870 Meeker was again absent for months when he traveled to Philadelphia and New York City on a trip to promote the prospects of Puget Sound to an eastern audience. During his hop years Meeker often traveled to New York and other eastern cities for extended stays, only occasionally bringing his wife with him. He made four trips to Europe in the hop years, accompanied by Eliza Jane just once. Meeker spent most of the years 1898–1901 in the Klondike, only coming home for a brief month or two. He was absent from May through October 1905 at the Lewis and Clark Exposition and again from February 1906 through January 1908 on the Monument Expedition. During these last three years, Meeker was with his wife just three months. These long absences were no doubt difficult. There are indications in his letters from the Klondike that she was lonely and wanted him to come home.

That he deeply loved Eliza Jane and that she was a trusted partner in his various enterprises (although, it seems, not always willingly or with the same tolerance for risk) is made clear both from his correspondence to his wife and his daughter Carrie, and from commentary by friends such as William Bonney and George Himes. It is only from Ezra's writings and brief mentions in the newspapers of the day that we learn of her active involvement in the family hop business; that she served by his side as Lady Commissioner to his role of Commissioner for the Washington State exhibit at the American Exposition at New Orleans in 1885–86; that from her home in Puyallup she ran the business of drying foodstuffs and shipping them north to her husband in the Yukon when he was operating the Log Cabin Grocery there; and that she sewed on his buttons, packed his bags, and raised their children. Her personal touch is just hinted at. "Jane 'slipped up' to the desk and quietly placed a nice bouquet upon the corner, as we were writing.... The house seemed a little more cheerful, the home more pleasant, because of the quiet act."[15]

The first indication of Eliza Jane's health problems came in 1903, and that was minor. Meeker wrote, "Mamma is real well but very forgetful."[16] Throughout 1904 and 1905 Meeker always referred to her as "well" in his correspondence. When he left home in late January 1906 to begin the Monument Expedition, the plan was for her to join him in Indianapolis where they would spend the winter. She remained behind, living again at their Puyallup home with Katherine Graham acting as a caretaker. As he traveled east through Oregon and Idaho he wrote to her often. In August, while in eastern Nebraska, Meeker received a letter from his daughter Carrie Osborne that contained alarming news about a serious change in Eliza Jane's health.

From that point on the correspondence between daughter and father make clear that Eliza Jane was suffering from some form of dementia. By November 1906 Eliza Jane had been moved to Seattle and was being cared for by Carrie.

Meeker struggled mightily with his conscience over the proper course of action. He was thousands of miles away from home and was earning money enough from the sale of his books and postcards not only to finance the Monument Expedition, but to help pay for his wife's care at home.

> And Mamma. How is Mamma? Is she now more contented? Miss Graham wrote she was much changed of late and probably would not know me if I were at home....I don't know what would become of me if I were at home. I know its cowardly to shrink from the inevitable, but Caddie, what more could I do for her comfort and happiness if I was at home.[17]
>
> This last letter again brings to the forefront as to where my duty lies. The question, can I be of any service to Mamma or relief to you by going home; If I can not, ought I not remain here to attend to my affairs and provide the means to help in the care of Mamma and avoid becoming a dependent myself in due time.[18]

Eventually Eliza Jane was moved into South Sound Sanitarium. When Meeker finally interrupted the Monument Expedition in February 1908 to take a month-long trip home to see his ailing wife, Eliza Jane no longer recognized him.

Accomplishments

What did the Monument Expedition actually accomplish? In meeting his first goal, marking the trail, Meeker was quite successful. He was present at the dedication or erection of fifteen markers in 1906. He ordered and had delivered six other stone markers, although he wasn't present when they were erected. He personally set twenty-five posts and painted rocks at many important sites along the trail. And within a few years twenty-four monuments went up at locations that Meeker recommended. Before he died in 1928 nearly two hundred markers lined the trail.

The second goal of perpetuating the memory of the pioneers who traveled that trail was also successful. The sale of 19,000 copies of Meeker's books was just the surface of his one-man educational

Meeker and Mardon at Huntington, Oregon, April 25, 1906. From *Pilgrim and Pioneer*

Huntington, Oregon, 2006. *P. Ziobron photograph*

campaign. Meeker also sold nearly a million postcards that depicted his ox team and prairie schooner and various scenes along the trail. He lectured to audiences from coast to coast. Virtually every town he passed through published a newspaper story about his expedition. He made contact with state historical societies and garnered their support for his project. The Sons and Daughters of the American Revolution signed on and began putting up their own markers. Meeker was seen or heard by hundreds of thousands of Americans during his two years on the road.

But perhaps his biggest impact was in the two cities that were the center of political power in the United States in 1906—New York and Washington, DC. Meeker made a splash in New York that no doubt exceeded his wildest hopes. He entered the city on August 17 and departed October 17. Hardly a day passed when he wasn't in the newspapers. Then as now, when a story made the New York papers, it was also quite likely to be in the national press as well. Meeker announced his presence by driving his ox team into the city without a permit and promptly got William Mardon arrested and the oxteam and wagon impounded by the New York City Police Department. After a long public fight with city hall, he was allowed to bring his "outfit" onto the city streets where he encamped for weeks. On September 19 he paraded his wagon down Broadway from Grant's Tomb to the Battery surrounded by immense throngs of cheering people. He even paraded his way across the Brooklyn Bridge.

The New York scene was familiar to Meeker. During his hop years he spent weeks and even months there negotiating with the heads of the railroads, shipping magnates and politicians—mingling with all varieties of the rich and famous. Meeker was quite comfortable doing so again in 1906, even though he had traded his business suit for the unkempt outfit of the old pioneer. One night he abandoned his camp and lodged for a time at the home of Robert Bruce, a leader of the National Highways Association (NHA). In 1922 when Meeker moved from Seattle to New York and needed an office for his Oregon Trail lobbying machine, it was the NHA that invited him to use their headquarters building, thus marrying Meeker's Pioneer Way project with the association. In enlisting wealthy and influential people to the cause, his network building was just beginning.

Meeker's first stop in Washington, DC, in November 1907 was at the White House for an audience with President Theodore Roosevelt who, before the day was out, endorsed Meeker's project. Meeker then stayed a month in the nation's capital, attempting to usher through Congress a bill that supported his trail goals with federal funding. House Bill 11722, the Humphrey Bill, was named for Washington State Congressman William E. Humphrey. It authorized President Roosevelt to appoint a commission to locate the exact route of the Oregon Trail and to erect monuments at appropriate locations. The project was to be funded by a $50,000 appropriation. Meeker testified in front of congressional committees and lobbied the Speaker of the House of Representatives, Joe Cannon. Although the bill did not reach President Roosevelt's desk, the power brokers in the United States were now well aware of Ezra Meeker and the Oregon Trail. If they hoped Meeker was simply a short-term nuisance, they were to be sadly disillusioned. What they could not have expected was the staying power of the now seventy-six-year-old champion of the Oregon Trail. He outlasted most of them, and he would be back in one fashion or another often over the next twenty-one years.

Even after tasting his first defeat, Meeker continued to build his network of connections and supporters. He wintered in 1907–8 in Pittsburgh under the care of another pioneer of 1852, Edward

Always a promoter, Meeker was not above superimposing images of himself and President Theodore Roosevelt onto this 1907 photo of the outfit in Washington, DC. *Author's collection*

Jay Allen. The two men had reconnected after a lapse of over fifty years. Allen had returned to Pennsylvania after a three-year sojourn in Washington Territory and became one of Pittsburgh's leading citizens. As the secretary of the Pacific and Atlantic Telegraph Company, soon to be known as Western Union, Allen's circle included names like Carnegie, Frick, and Mellon. Allen agreed, as did Robert Bruce and President Roosevelt before him, to help with Meeker's effort save the Oregon Trail. Meeker was building a network for the future. It would not be until the 1920s that it all came together, but the start was made with the 1907–8 Old Oregon Trail Monument Expedition.

BUILDING A COVERED WAGON

The covered wagon (or prairie schooner) Meeker traveled with on his various treks around the country between 1906 and 1915 was a new wagon constructed specifically for the adventure.[19] At the end of his wanderings Meeker donated the wagon and his stuffed oxen, Dave and Dandy, first to the Tacoma Metropolitan Park Department, and later to the Washington State Historical Society. For decades they were on display in a glass case in the old Ferry Museum building in the north end of Tacoma, Washington. That building today is the Washington State Historical Society Research Center. In 1996, when the historical society acquired a new home for its museum on Pacific Avenue in Tacoma, the wagon and oxen were moved to that site. A few years later the wagon was put into storage in the basement of the research center and the oxen became part of a hands-on display about transportation in the museum. In 2011–12 the wagon was brought out of storage for a time as part of a museum display titled "Icons of Washington."

According to Meeker the wagon he used in 1906 was constructed from parts of three different wagons. "[M]y wagon is just the thing—is made up from three wagons (the iron) that crossed the plains in the early fifties," he wrote to George Himes that year.[20] It was a great point of pride for Meeker, and in keeping with the significance he bestowed on this historic mission, that at least portions of the new wagon should incorporate authentic pioneer parts of the old. Meeker's 1905 journal tells us that (at least one of) the wheel hubs and iron work came from a covered wagon belonging to Rudolph Giesy that was built in Bethel, Missouri, in 1853 and came over the Oregon Trail in 1855.[21]

Rudolph Giesy was a member of a Bethel, Missouri, religious and communal colony that traveled west in 1855, eventually ending up in Aurora, Oregon, in the Willamette Valley near present-day Salem. Giesy never married, but apparently thrived in his new home. When Giesy died in 1895, his wagon eventually came into the hands of a man named Brilis in Pacific County, and it was from Brilis that Meeker obtained the wagon parts.

Meeker had a long connection with the Giesy family starting around 1857, when Meeker's name appeared on the list along with Christian Giesy as a member of "The Geographical and Statistical Society of Washington Territory." At the height of his prosperity as a grower and broker of hops, Meeker's 1895 business records show him doing business with various Giesys.[22] In October 1905 Meeker was aware that Rudolph Giesy's wagon, or parts of it, still survived, and he sent for them.

Stenciled under the hinged seat lids of Meeker's wagon is "Crane and Wyllys Makers Puyallup, WA."[23] Crane, who was paid $30.15 for additional wagon parts,[24] was probably Richard E. Crane who was listed in the 1900 census as a carriage maker living in Puyallup. Wyllys was likely Charles Wyllys who was listed in the 1900 census as blacksmith, also living in Puyallup. Meeker supplied the following confirmation. "I think Wyllis [sic] who ironed my wagon is yet in Puyallup; Crane who made the woodwork has died. He selected the timber in one of the houses doing business in Tacoma."[25]

The 1906 blacksmith, Mr. Wyllys, came under some serious criticism from Meeker when he ran into a problem with the wagon brakes. "[M]y wagon is nearly finished but the mush headed blacksmith has forgotten how to make that brake although I had him go and examine it at the Fair."[26]

From 1906 through 1908 Meeker's wagon (with the 1855 Giesy hub) traveled from Puyallup, Washington, to Washington, DC, and back. In the spring and summer of 1909 it was exhibited in Seattle at the Alaska-Yukon-Pacific Exposition. In the fall and winter of 1909 it traveled through southern California and participated in the Pasadena Rose Parade on New Year's Day 1910. In 1910–12 it made another trip across the country following the Oregon Trail to Kansas City, Missouri, and east as far as Niagara Falls, finally venturing south to San Antonio where it was displayed for a time at the Alamo. In June 1914 it made a final journey south to San Francisco where it was displayed in 1915 at the Panama-Pacific International Exposition. In 1916 it found a home with the Washington State Historical Society, where it has resided ever since.

Meeker's wagon stored in the basement of the Washington State Historical Society Research Center in Tacoma, Washington. *P. Ziobron photograph*

CHAPTER 4

Disaster at the Alaska-Yukon-Pacific Exposition

Ezra Meeker was a risk taker. He had been all his life. Sometimes it worked out, and sometimes it didn't. The bad outcomes cost him and those around him dearly. He was rewarded beyond all expectations by his decision to grow hops in the 1860s, but then he failed to heed his own advice to diversify. Additional decisions in the 1890s to embark on several financially large projects at the same time left him dangerously overextended when the Panic of 1893 hit. This was ultimately responsible for his bankruptcy in 1896 and the financial ruin of his family.[1] In 1909 he took another gamble on the Alaska-Yukon-Pacific Exposition, one he was warned not to take, and his monumental miscalculation led to another financial disaster—one that would test his fortitude and character deeply. A self-assurance that bordered on arrogance dug the hole. His refusal to quit helped him to climb out, but it took years.

Germination

Late in 1907 in Washington, DC, at the end of his long journey across the continent, Meeker heard a rumor that intrigued him. Plans were afoot to hold an exposition in Seattle. On Christmas Day 1907 he asked his daughter, Carrie Osborne of Seattle, to "inquire about the prospects of holding that Alaska exposition in 1909; if one is to be had, I want to get a good location for my team where I can sell my literature."[2]

Then came a chance encounter. Also in the nation's capital that winter was John H. McGraw, who had been the governor of Washington from 1893 to 1897. McGraw was one of the future exposition's vice presidents, and he was in Washington, DC, to ask Congress to fund a government exhibit at the upcoming fair. Somewhere the two met. McGraw urged Meeker to write Exposition Director General Ira A. Nadeau about his desire to participate. Meeker did so immediately, asking for a concession for an exhibit about the "Old Oregon Trail" and "Pioneer Life in the Old Oregon Country." He thought three acres would be adequate and filled a page outlining the possibilities of such an exhibit.[3]

Director Nadeau wasn't as sanguine as the governor. He told Meeker he would be glad to help him, but he needed to understand that the exposition would take a percentage of the gross receipts of any concession granted. Nadeau also said three acres was much too large. Then he offered a word of warning that Meeker totally ignored. "Such an illustration would be exceedingly interesting to me, but whether it would strike the average attendant at an exposition where noise, music and crowds are the attraction rather than such quiet reproductions as you contemplate, is a question."[4]

Meeker dismissed Nadeau's warning. He wrote back that he contemplated "both noise and music" in his pioneer exhibit.[5] In February 1908 Ezra traveled by train from Pittsburgh to Seattle, primarily to see his ailing wife. While home he paid a visit to Director of Concessions A. W. Lewis, and discussed his plan for an exhibit at the exposition.[6]

There matters rested. Meeker rejoined his outfit on the road in Missouri, heading west through the last months of the Monument Expedition. In June he was back in Portland and began a grand drive with his ox team and covered wagon north to Seattle. The press and the public greeted him as a returning hero. More to the point, he was arriving home with a good deal of money, funds earned selling his book and postcards—some $3,000 (around $75,000 in today's currency), and ready to spend it for the cause.

Late in the summer Meeker was informed that he had been granted a space at the exposition to set up his exhibit free of charge and had been given permission to sell his books. However, Ezra was told that he could not sell postal cards as that concession had been granted to the Portland Post Card Company; if he made some kind of arrangement with them the exposition management would have no objections. He was invited to come out to the exposition grounds and select a space for his exhibit.[7]

Meeker vs. China

Meeker immediately began preparations for his exhibit. Then came trouble. The possibility arose of China entering an exhibit, and the directors considered locating it in the space already promised to Meeker. Ezra wrote to Nadeau complaining that Director Allen refused to set the stakes for space he had been promised. Meeker had already sent advertisements to many newspapers along the line of his march of 1906–08, and he complained he could not in good conscience continue to solicit financing for his exhibit as long as there was doubt as to its location. He went on to outline the projected revenues that the exposition might expect to receive from his exhibit. He ended the letter with the following:

> I know there will be a storm of protests if an attempt is now made to defeat the occupancy of the site designated by you for the Pioneers. Now don't you think the Pioneers are entitled to the best? It was this class that made it possible to hold this great exposition by the present generation…We are not willing to go way back and sit down. We are entitled to a front seat.[8]

In late November 1908 the directors expressed their regrets to Meeker and suggested that all parties involved let the matter rest for a week or two. Meeker's answer was quick and brief, "I consider you have already assigned a site for my exhibit and I am unwilling to change unless another is given me which in my judgment is as good as that already agreed upon."[9] An alternate site along the shore of Lake Washington was not acceptable.

By the end of the year Meeker agreed to place the matter before a committee that was empowered to settle disputes. On January 2, 1909, he wrote to George Himes and said if the Chinese ended up kicking the Pioneers out he would, "expect to leave the state for good and all."[10] Meeker wrote former Governor John H. McGraw asking for a meeting and renewing his threat.[11]

There are no further letters in the Meeker papers about the dispute. It seems Meeker was satisfied with the final decision, as he didn't leave the state, and he went forward with his exhibit. Ezra settled for a site in the Pay Streak (the midway) instead of next to the Washington State Building as he was originally promised. As things turned out it might have been better, at least financially, if he had packed his bags and departed.

The Exhibit

Meeker's AYPE exhibit was essentially divided into two very different parts—the "Pioneers Restaurant" and the "Oregon Trail and Pioneer exhibits." The restaurant was built in a log cabin style. It was a large structure staffed by thirty-two workers and it fed a lot of people. On Seattle Day some two thousand customers sat down at its tables. Over its lifetime the Pioneers Restaurant grossed $3,000 and barely met its expenses. Many of the staff members were relatives. The use of family members in his various projects and expecting them to enthusiastically support him would eventually cause a breach with some of them. It did not go well by the conclusion of the AYPE either when he found himself unable to pay the wages that some of them were owed. Relatives included Aaron Lincoln Meeker, Ezra's half-brother, who lived near Okanogan, and his family; John Valentine Meeker's daughter, Mary Frances Bean, and daughter Vida Bean; Meeker's half-sister Malinda Meeker Daniell and her two daughters from Omak; granddaughter Bertha Templeton; and daughter-in-law Clara Meeker. Meeker's idea was to lease the Osborne home in the Madison Park district of Seattle to use as a dormitory for these family helpers. He assumed the Osborne family would move across Lake Washington to their summer cabin during the run of the exposition. Meeker doesn't mention the Osbornes' reaction to his plan to take over their home for the four months that the AYPE would run.

The Oregon Trail exhibit consisted of several buildings and objects that were intended to demonstrate the hardships of the trail and pioneer life, and it included a replica Indian encampment. Admission to this area was ten cents. The main attraction of the Oregon Trail exhibit, of course, was to be the ox team and covered wagon, from which Meeker hoped to sell many copies of his books, *Pioneer Reminiscences*, *The Ox Team*, and the newly printed *Ventures and Adventures of Ezra Meeker*. William and Cora Mardon were in charge of Dave and Dandy, the wagon, and selling books.

Other exhibits included a replica of the Jason Lee 1834 cabin near Salem, Oregon. This was the first Methodist Mission House in Oregon. Meeker's older brother, John Valentine, age eighty-five, served as docent for this exhibit.

Housed in a replica pioneer cabin was a demonstration of spinning and weaving. The plan was to make products in the cabin that could

Overview of Meeker's exhibit at AYPE, restaurant in foreground, pioneer exhibits behind. *Ventures and Adventures*

Meeker's restaurant at AYPE, *Ventures and Adventures*

be sold to the watching public. Many letters and some cash ($200) went back and forth between Meeker and Mrs. Jennie Lester Hill, the Superintendent of the Berea College of Domestic Science & Fireside Industries in Berea, Kentucky. Mrs. Hill had access to old-fashioned spinning wheels and looms and could send two students who knew how to use them. Meeker hired the students, Miss Cruch and Mrs. Hays, for forty dollars monthly and meals, and paid their train fare from Kentucky. By the end of April he informed the Kentucky ladies to ship the spinning wheels, looms, and materials. The cabin was ready.

An original blockhouse stood on the crest of a man-made hill overlooking the pioneer village and restaurant. Meeker first tried to get the Borst blockhouse that stood along the Chehalis River in Centralia, Washington, and which, incidentally, is still standing there today. He tried next for the Ebey Landing blockhouse on Whidbey Island. Failing there also, he was finally given permission to dismantle and move to the exposition grounds a blockhouse that stood four and a half miles from Coupeville, Washington, on Whidbey Island.

One unusual display consisted of a rifle embedded in a section of tree trunk. Jacob Price of Grand Mound loaned the tree and gun to Meeker. Meeker wrote to George Himes, "I am getting the oak tree grown on [Grand] Mound Prairie with the gun in the forks. The parties shipping it think you have the history of it."[12] Its history was thus: at the outbreak of the Puget Sound Indian War in 1854 the pioneer owner of the gun and his Indian neighbors agreed that they should remain friends, and as a token of mutual confidence the gun was placed in the fork of the tree as an emblem of peace. There it remained for fifty-four years as the tree grew around the weapon.

Financing

There are few details relating to the construction of Meeker's AYPE exhibit in the Meeker Papers. Berger Brothers of Seattle were the architects. The lumber and flooring mostly came from the Ranning Lumber Company. Charles Hood, Meeker's Puyallup friend, and his assistants did the plumbing. Meeker signed contracts with local merchants to deliver daily all the milk fourteen cows could produce, three gallons of cream, a case of fresh eggs, and three cases of strawberries. Beyond this his letters are silent. What we can infer is that the undertaking was both huge and expensive.

Meeker came home from the Old Oregon Trail Monument Expedition in 1906 with approximately $3,000. In a January 1909 letter he described to his son his hopes for that money, especially regarding the care of his wife, Eliza Jane, who was in late stage dementia. She had been placed in the South Sound Sanitarium after she was found meddling with the stove in her room. The risk of a conflagration was too great to leave her unattended. Carrie or Ezra visited her nearly every day at the sanitarium and the Osbornes began installing a basement furnace that would heat the whole house, thus removing the stove in Eliza Jane's room and its dangers. This would allow the family to move Eliza Jane out of the sanitarium and back into the Osborne's home. Meeker told his son, "I will have to keep a lady trained nurse constantly with her. I feel rejoiced to know that I will be able to do this for several years even if I make no more accumulations, which I hope to do this summer at the Exposition."[13]

By the end of March Meeker had solicited $1,000 from the businessmen and citizens of Puyallup to support the AYPE exhibit, and he asked William H. Paulhamus to take the lead to secure a like amount from the city of Sumner.[14] By April 4 Meeker had poured much of his own $3,000 and all he had collected elsewhere into the project, and he needed $1,200 more to complete the construction of all the buildings. He turned to his old friend Louise Ackerson for a $1,000 loan to complete the project. Mrs. Ackerson insisted on having the note cosigned. Much to their everlasting regret, his grandson Joe Templeton, son-in-law Eben Osborne, and daughter Carrie Osborne cosigned the note. The drama concerning this loan would play out all the way to 1912.

THE EXPOSITION OPENS

On June 1, 1909, the Alaska-Yukon-Pacific Exposition opened its doors to the public. Meeker listed his earnings for the first few days of the Exposition. They were not what he hoped for—only $220.38 for the Oregon Trail exhibits. By July gate receipts were so low that Ezra installed a carnival ride on the site called the Joy Wheel. This did much better. On Seattle Day in September, 7,000 people paid twenty-five cents each to ride the wheel. In fact, the ride was so popular that the exposition directors tried to install a second one elsewhere on the exposition grounds, much to Ezra's ire. Meeker warned the directors not to take this step and was successful in preventing an unwanted competition.

The $1,750 income brought in by the Joy Wheel on Seattle Day should have been enough to raise Meeker's spirits. Unfortunately Meeker was only a 10 percent partner in the "wheel" so his take on this day was $175. Most days he earned much less, as little as fifteen to twenty-five dollars a day. The restaurant did better.

The Exposition Closes—for Meeker

By August 17, 1909, Meeker decided to close the Oregon Trail part of the exhibit. It was simply losing too much money. He wrote Mrs. Jennie Lester Hill and told her the Kentucky college students were starting home the next day and asked her what to do with the looms and spinning wheels and the remaining products they had made. He promised to send the agreed upon money as soon as possible. Two weeks later he wrote Mrs. Hill, again promising to pay what was due as soon as possible. He told her about his financial disaster concluding, "Better luck next time as has been well said, and so say I even if it does not come. It was all the money I had."[15]

Realizing that the ox team and wagon were now his only source of potential income, Ezra sent William and Cora Mardon with the outfit to Portland and then to the state fair being held at Salem, Oregon. They continued on to Eugene and then eventually to California. Meeker stayed behind to operate the restaurant. He also installed a second carnival ride on his site despite having to fight the exposition management over their share of the receipts.

Meeker obtained permission from the mayor of Portland for the Mardons to sell literature from the wagon on the city streets.[16] By September 2 he was getting anxious to hear from them as to how things had gone, and he suggested that the couple might find "a good field in the hop yards near Independence [Oregon] if you get there in time."[17] He was pleased to learn that they sold $22.25 worth of books and postcards their first day in Oregon. He asked them to write daily, if only a postcard. On September 10 he wrote, "I am draining my bank account up close…. Now if perchance you do make some sales send it along even if it's but a small amount. Attendance yesterday 15,000; receipts in restaurant $70.00—how is that for high or rather low."[18]

On September 15 Meeker wrote the law firm of McCrea & Branigan explaining his situation and laying out his plan for paying off his creditors. He estimated the restaurant was earning about fifty dollars daily and the Joy Wheel fifteen. A week later he wrote Jacob Price of Grand Mound. "My business has failed; my money all gone. My property under attachment and I can't do a thing. So you will have to come down and move the tree or send money to pay for moving it."[19]

Meeker left the exposition grounds and his creditors near the end of September. He took the train south to Oregon, stopping briefly in Halsey to visit his daughter Ella. He arrived in Salem on September 29 to find Mardon sick in bed with what the doctor feared was typhoid fever. There were almost no books left in the wagon to sell as Mardon had shipped most of them ahead to Eugene. His purpose in doing this was to lighten his load, as he had expected to drive the team there before he fell ill. Meeker also learned at the train depot that he could ship the oxen and wagon directly to San Francisco for what seemed like an incredibly low price of $40.35. His fare, if he rode in the car with the oxen, was just fifteen dollars. The trip would be four days. Meeker jumped at the opportunity, unsure whether Mardon could follow. Meeker fired a letter off to Willis Calder of the Rainier Printing Company asking him to ship a couple of boxes of books (*Ventures and Adventures*) to San Francisco as fast as possible as he wasn't sure he could get the books at Eugene forwarded promptly.

He left for San Francisco at 3 a.m. October 1 in a boxcar with the wagon, oxen, and his dog Jim. He set up his cot, made his bed, and settled in for the four or five day trip.[20] He arrived in Oakland just before dark October 6. Along the way he sold fourteen dollars' worth of books and postcards. The next day, quite relieved, he wrote his daughter, "Mardon has arrived and is better; Mrs. M. also here."[21] He wrote the attorney for one of his creditors, "[T]hey must judge for themselves as I will eventually pay whether they begin a suit or not."[22]

Meeker had signed up for San Francisco's Portola Day Parade and Carnival, a five day celebration of the city's recovery from the great earthquake and fire of 1906, and had been given what he considered a fine location on the carnival grounds. But he had no stock of books in San Francisco and the city authorities did not want him on the streets until the start of the carnival on October 16. All he could do was wait. While biding his time he went to the Unitarian Church in Oakland to visit Mr. Simonds,[23] his former Seattle pastor. Ezra received an invitation to address the study class and to meet some influential men, some of whom were old timers. Among them was the mayor of Oakland who held the key to letting Meeker sell his literature in that city.

Sad News

On Saturday, October 9, 1909, Eliza Jane Meeker died at the South Sound Sanitarium in Seattle. It took the family three days to locate Ezra. They thought he was somewhere on the streets of San Francisco with his ox team. Instead he was in Oakland visiting with Mr. Simonds. Only after he moved the outfit to the carnival grounds and set up his tent did he receive the sad news of Eliza's death. That night, Monday, October 11, Meeker started the long train ride to Puyallup.

A mere three and a half hours after his train pulled into the station at Puyallup the family and some close friends gathered in the parlor of the old mansion, now owned by Carrie. At 11 a.m. on October 13 the city of Puyallup, as a measure of respect, closed all the stores in town. A grandson, Reverend Harry Templeton of Vancouver, Washington, conducted the services. Daughters Carrie and Ella from Seattle attended, but the Meeker's other two surviving children were absent. First-born son, Marion, lived in Redlands, California, and the youngest daughter, Olive, lived in Eagle, Alaska. The *Puyallup Republican* said, "The services were private and everything was in keeping with the simple quiet life of this pioneer woman. Interment was in Oakwood cemetery."[24] Meeker's pocket calendar for the day reads, "Wednesday October 13th Arrived in Puyallup 7:30 A.M. Buried my wife."

Ezra wrote to Katherine Graham,

> After a simple service in the parlor at the old house she loved so well, we laid her to rest on that beautiful spot on the hill....
>
> Although I had known the end would come soon and could not be delayed for long, yet after all I was not prepared; I could not realize she was dead. She never looked more beautiful in life than she did in death, with her placid countenance as though she might awaken and smile on us again and when I turned from the grave and knew; I then felt as I had never felt before, not only a willingness, but a longing to join her in her new and beautiful home. It is well for me that I have plenty to occupy my mind and that my health is good.[25]

The next morning Meeker started back for San Francisco. He arrived as the carnival opened. Meeker wrote to his daughter some thoughts about the funeral. "Now that it is over, I am glad I made the trip. I now feel more reconciled than I would had I not been there... yes, I am indeed rejoiced to have made the trip and feel I never would have been satisfied had I not gone."[26]

Waiting for him in San Francisco was another heart-wrenching letter:

> Dear brother
> Only three more days of the A. Y. P. E. We have expected to hear from you for some time but have not yet. Jessie [Meeker's niece] has been sick in bed for almost a month and has talked and worried over her wages so much I thought I would write to see if you had forgotten you owed her $24 as she said you thought you could send it the next week after we left but it never came. She wrote from Wenatchee for it but did not get it, she borrowed money in Wenatchee to get home and it has not been paid yet. I too need my money but I am willing to wait if you would only send Jessie hers for she has suffered so much and still suffers, she feels hurt too to think you paid Olive & Beulah all they earned and did not pay her.
> She said she worked just as hard and earned hers as well as they—I hope you will think this over and if it is in your heart to pay the child do so at once for what worries her worries me for she is very dear to me and she has suffered so with reticular rheumatism and nothing seems to cure her. I am pretty well tired out caring for Jessie night & day. So I will close hoping to hear from you soon. I am
> Sister Fanny G. Meeker [Aaron Meeker's wife][27]

Ezra Meeker, for years one of the wealthiest men in Washington Territory, was as broke as he had ever been in his life. He simply had no money to pay any bill, even such a just and modest one from his ailing niece. Meeker would spend the next four months in California trying to fix this situation.

CHAPTER 5

Retreat to California

To pay off the debts he owed to his niece Jessie, Louise Ackerson, Rainier Printing Co., and countless other creditors he left behind when fleeing Seattle, Meeker needed California to be kind to him. It had to offer more than sunshine and a gentle climate. Meeker needed customers willing to purchase his books and postcards, and this, of course, meant securing permits that allowed him to sell on city streets.

When Meeker finally arrived on the ground in California he sent out letters to the mayors of major cities including Stockton and San Francisco asking for that permission. In this effort he had mixed success. Some cities opened their streets to him. Others gave him a gentle no. Some, like Oakland, delivered a more emphatic no. When this happened he was usually forced to set up shop on a vacant lot, or was confined to a specific street corner for a certain period of time. Local fairs, city celebrations, and parades offered a way around these restrictions. Meeker took every opportunity possible to place himself in front of the "people," but problems with permits would plague him throughout his California stay.

On October 21 Meeker and the Mardons drove the wagon in San Francisco's Portola Parade, for which they received twenty dollars. He described the turnout as massive. "I never saw so many people in one day in all my life. Mrs. Mardon and I went out after a hurried afternoon meal and sold twenty-three dollars but the people were not in the mood to buy anything but confetti and rattle cow bells; however I think we did pretty well."[1] He concluded the letter asking his daughter to telephone the Rainier Printing Company and ask them to hurry up with a shipment of books.

Meeker's inventory of books on hand was low, and he faced unexpected and vexing problems resupplying. Replenishing his stock promptly would be a recurring problem for the next two years. For one reason or another (possibly a large unpaid bill), Mr. Calder at Rainier Printing could never seem to get them shipped as fast as they were needed. Meeker's original plan had Stockton as the next stop after San Francisco. The Mayor of Stockton had granted selling privileges as requested, but due to a mixup about dates, Meeker chose to pass up the "Rush of 49" celebration in that city. Instead he went to Oakland. It was a decision he came to regret.

Oakland authorities restricted him to the corner of 15th and San Pablo Avenue, and for only a short time. He tried to change that, but failed, even after making a personal request of the mayor. The unhappy pioneer instead secured permission from the YMCA to camp on their lot at 14th and Jefferson Street for one week beginning October 28. Meeker again attempted to plead his case to city hall but Mayor Mott refused to see him. A passionate letter full of references failed to move the mayor. Discouraged at how slow sales were going Ezra wrote his grandson, Joe Templeton, a Seattle attorney on whom he had inflicted the onerous task of dealing with his AYPE creditors. Templeton handled his grandfather's affairs mostly without pay, doing him a huge favor that Ezra did not fully appreciate. Joe Templeton and Carrie Osborne acted as his Seattle liaisons for the next three years, always on-call for the next instructions and seemingly willing to do any number of tasks requested, often to their own financial detriment, and all mostly without complaint. Meeker summed up his situation and introduced a famous acquaintance, Joaquin Miller, the renowned "Poet of the Sierras."

> I had hoped for better results here but have met with obstacles at the city hall, but this morning received a splendid notice by one of the papers and the effect is visible at once. I can do no otherwise that I can see, than to hammer away the best I can for I see no other field more promising than that outlined before I left home....I have spent the afternoon and part of the evening with Jacques Miller, the poet and had dinner with him at his quaint abode out in the hills 4 miles from Berkeley. He treated me extremely cordial and will write an open letter commending my work, which I doubt not will be helpful; he is a strange make up indeed."[2]

In November Meeker sent Templeton a money order for one hundred dollars to begin payments on his notes. He told Joe, "The signs of the times is more encouraging and I think it's going to work out all right now that I have escaped from that 'Sodom & Gomorah (how do you spell it?)'."³ He was referring to Oakland. Meeker and the Mardons were well received in Alameda and Hayward where they were allowed to set up their tent and camp on the city streets. From Irvington he sent an upbeat progress report to his daughter that included a comment about Mrs. Mardon's cooking. "[T]he table fare is good—too good—and I am comfortable and well."⁴

SAN JOSE TO LOS ANGELES

San Jose was also kind to Meeker. The mayor, Charles W. Davison, gladly gave him permission to camp and sell on the city streets. The local press covered him favorably and the local Federated Women's Club heartily endorsed his work. He lectured to the ladies and then to an audience of six hundred at the state Normal School. At church, the pastor called attention to his presence, resulting in what he called a "nice reception." Ezra was also invited to dinner with Herbert Bashford, author, poet, and book reviewer for the *San Francisco Bulletin*. Most gratifyingly, the books finally began to sell, bringing in over a hundred dollars in three days just from sales at his camp. Meeker sent Templeton another remittance to pay toward the Calder bill, keeping back a hundred dollars for emergencies.

He next headed south, telling his daughter-in-law Clara Meeker, "I have definitely determined to ship from here to Los Angeles and get out of the rain belt."⁵ The team was shipped with the Mardons by slow freight while Ezra went ahead to "spy out the land." His rent-free camp was set up on a vacant lot on the corner of 8th and Broadway, opposite the largest department store in the city, one floor of which contained the public library and another a theater. It was one of the busier areas of the city, putting Meeker's outfit in contact with large numbers of people. The move from San Jose to Los Angeles cost seventy-five dollars.

The Los Angeles City Council gave him a thirty-day permit to sell from the Broadway lot. Meeker wasn't certain he would remain that long. He told his daughter, "[W]e have a much larger percentage of 'Rubber neckers' than in the east and I think the sales are going to be moderate."⁶ He was correct. Sales went in spurts. One day they would reach the forty-dollar mark. The next, "I verily believe 2000 people visited my camp yesterday to ask all sorts of silly questions and bought—shall I write it? Ten dollars, cards and books."⁷ By December 3 Meeker was down to his last box of books. He fretted about the possibility of facing the Christmas season without any stock to sell, but concluded to remain where he was until the week-long Los Angeles International Air Meet began on January 10.[8]

Gradually word of mouth brought customers, and the promise of Christmas sales loomed ahead. He wrote his daughter that camp life was pleasant, that Mrs. Mardon was a "good clever cook" who took care of his clothes and other housekeeping chores, and that he had no complaints.[9]

Ezra's December letters further described camp life in southern California. He reported that he had replaced the wagon cover and that he had a new stove, "a better one with a better oven so now we can have our toast without burning and baked apples too….We do have toast browned just right and coffee (made in one of the new kind of coffee pots that boil up in the center through a tube) that is exactly right and the baked apples are delicious—what more do I want."[10] In another letter he offered a complaint about Cora Mardon saying she "talks too loud at times and in what seems in an angry tone (not to me though) so you see I am not without my household war…but I sometimes call [sic] her down in a quiet way aside that I can see is doing her good."[11]

Ezra took his comfort where he could.

> The faithful dog Jim lies at my feet asleep and lies at the foot of my bed every night and wakes me up every morning when daylight comes. I have often wished he could talk—he does with his eyes and some of his movements for he seems to understand every word I say; he has been my constant companion ever since we drove out of the front door yard January 29th, 1906 and he has had more adventures than I could recount in many pages of writing.[12]

For the Thanksgiving holiday he left the outfit in Los Angeles under the care of the Mardons and made a trip to Redlands to see his son Marion and family. Ezra found everyone well but passed up a Christmas invitation despite Marion's promise to "have the same kind of turkey as on Thanksgiving."

Christmas was coming and Meeker's thoughts, never straying far from the urgent need to repay his debts, were scheming afresh. The Puyallup property had a number of holly trees on it and Ezra saw a potential market in Los Angeles. He asked Carrie to send him some samples along with estimates as to what they should sell for. He told her to consult Charles Hood of Puyallup on the matter. Ezra assured her that Marion's daughters could sell a lot of wreaths "here in town of 12,000 inhabitants where there are 26 millionaires."[13]

Meeker then wrote his grandchildren Joe, Frank, Harry, and Lloyd Templeton early in December, asking each in turn to send him one or two wagon loads of Christmas trees to sell at his lot in Los Angeles. An early northwest storm deposited eight inches of snow making the roads impassable and the Christmas tree plan died. So too did the plan to sell the holly that grew on the mansion grounds. Meeker usually sold it each winter to florists around the northwest, earning around $600, but his hopes to sell holly in Los Angeles were dashed. Carrie had given it all to Charles Hood as a thank you for his help putting in the plumbing at the AYPE. Charles had little holly left, and what there was wouldn't get to Meeker in time to sell before Christmas.

A Decision

Just before Christmas Meeker made an important decision. He told Joe Templeton that he planned to drive to Pasadena where he had been granted a permit by the city council, and that he would be participating in the Rose Parade. However, it was now abundantly clear that he could not pay off his debts from a campaign in California. He would have to go out again on the Oregon Trail where he knew "the people will turn out almost in mass and buy freely." Another Oregon Trail bill would be introduced in the upcoming Congress, but Meeker expressed doubts that it would pass. He told Templeton, "If my good health continues, I have faith to believe I will not need to go far out on the Trail before selling enough to clear off the book account and secured notes."[14]

On December 26, Meeker wrote his daughter that he had sent Templeton $350 to pay on the printing bill. He also told her of his decision to make another trip over the Oregon Trail.

> Of course, as you know without telling I also have as a motive, the selling of my book and provide for paying off my debts.... The Mardons will go with me. Mardon is a much better man than when I first knew him. Mrs. M. in adversity is more reasonable in her demeanor, is a good cook, saving and industrious. ...Both are anxious to make the trip. I will drive to Pasadena tomorrow and remain there for the week and go into the flower show parade [Pasadena Rose Parade] Saturday.[15]

54   Saving the Oregon Trail

Meeker, in fact, requested that Washington State Congressman William Humphrey and Senator Samuel Piles introduce an Oregon Trail Bill once again in the upcoming session of Congress, and he offered to come to Washington, DC, to testify on their behalf if they thought his testimony was needed. To Clarence Bagley he wrote,

> This last is rather a bold bluff for as a matter of fact unless I can do better than heretofore in California, I absolutely would not have the money unless friends of the work will help. I really do not believe the bill can be passed at the first session of the next Congress if the tariff bill shows up a deficit, but it's a good move to get the reports and the bill on the calendar of both houses.[16]

Pasadena—The Rose Parade

On December 29 Meeker sent Joe Templeton another hundred dollars on the Calder account, making $450 in all. He hoped by the end of 1910 to "be 'on my feet' financially and hopeful of at the same time to make substantial progress on the work of monumenting the Trail."[17] Ezra celebrated his 79th birthday camped in Pasadena. On Saturday January 1, 1910, Meeker and the "outfit" marched in the Rose Parade and won a first prize of twenty dollars. He used the money to buy a new suit and then moved back to his camp in Los Angeles for a week while he waited for the aviation show to begin. He secured a permit to camp on

Meeker at the January 1910 International Air Meet held in present-day Carson, California. Mrs. Mardon in wagon. *Author's collection*

the aviation grounds for ten days and thought he would stay throughout the meet if sales warranted.[18] The first two days were plagued with rain. Nevertheless, his camp was thronged with visitors and when the sun returned sales picked up. Meeker was fascinated by the airplanes. "I moved out here Monday [Jan. 10th] and have been busy watching airships from my tent soaring around the field…wonder of wonders."[19]

In 1924, nearly a decade later, Meeker would take his famous airplane flight, crossing the continent all the way to Washington, DC, advertising the Oregon Trail. That was in the far future, however. The next stop after the aviation meet was Long Beach, where he camped for three or four days with his ox team outfit on a vacant lot at the foot of Pine Avenue. From here Ezra sent another $200 to Joe Templeton with this instruction, "[M]aybe its just as well to clean up the Marion & Blake note and pay over remainder to Calder."[20] This was his first mention of bills from the AYPE other than the debts to Calder and Mrs. Ackerson. It also brings into better focus the extent of the financial disaster the exposition was for Meeker.

Marion came to the last day of the air show and went on to Long Beach for a four-day visit with his father. Also some old acquaintances came calling. "I have had a regular round of visiting with an old acquaintance—way back home, a lady who was at my wedding; another Ling Ballard,[21] and scores I had met in the east. I have set the 10th of March as the date to start out from The Dalles but I expect to be home by the third and remain with you two days."[22]

The trudge through California continued with a stop at Santa Ana, twenty-two miles away, followed by Pomona, Riverside, San Bernardino, and then Redlands. Meeker went ahead each day by rail to make camping arrangements while the Mardons drove the ox team and wagon, all the while selling literature. He wrote, "[T]he aviation cards like that I sent you sell rapidly."[23] On Feb 6 Meeker reported,

> I left the team last evening at Ontario about 30 minutes down the line and came here to spend the Sunday with the Folks [Marion and family] and will meet the team at Riverside Tuesday [Feb. 8]; I have been well received on the way here with better sales than at first but the towns are small. If I had a couple hundred more books would have stayed until the 1st but after all its about as well to be out early on the Trail; I am not having any trouble with the city authorities—they let me camp where I please and sell where I please and of course doing better.[24]

Meeker ended his California stay with a weeklong visit with Marion's family at Redlands. He arrived home in Seattle on February 22, 1910. The Mardons and the outfit trailed behind in a freight car. Their destination was The Dalles, Oregon. Still unpaid was much of the Rainier Printing Company bill and Mrs. Ackerson's entire note. But the Oregon Trail was calling, and with it the hope of better times.

Meeker's camp on Euclid Avenue, Ontario, California, February 6, 1910. Cora Mardon is at the table selling literature. William Mardon is standing by the oxen. Meeker is by the wagon reaching up toward the collie Jim. *Author's collection*

CHAPTER 6

The Second Old Oregon Trail Expedition, 1910

By most reckonings, Ezra Meeker's 1906–8 Old Oregon Trail Monument Expedition would be considered a smashing success. Meeker turned what was originally an old man's passion to mark and save the Oregon Trail into a national effort. He traveled successfully by ox team from Puyallup, Washington, to Washington, DC, and back—3,650 miles according to the banner attached to his wagon. He drove the team down Broadway in New York City from Grant's Tomb to the Battery drawing crowds of thousands; drove up Pennsylvania Avenue to the White House where he had a meeting with President Theodore Roosevelt and convinced him to sign on to the effort to preserve the trail; got Congress into the act with the introduction of the Humphrey Bill; wrote a book about the expedition that went through four editions and sold 19,000 copies; sold hundreds of thousands of postcards from the wagon to meet expenses; lined the trail with some twenty monuments; and came home with a good amount of change in his pockets.

The second expedition from 1910 to 1912 could best be described as an ordeal. Ezra himself called it a failure. The "particulars" of this story have never been told. The second expedition is in its own way as remarkable as the first expedition, perhaps more so because Meeker, who would turn eighty years old on the journey, was still carrying on like he was a young man. Ezra kept a journal only for the first year. Other "particulars" were ferreted out from over one thousand letters covering these years found in the Meeker Papers at the Washington State Historical Society Research Center in Tacoma, Washington.

Everything about the second expedition was hard—from beginning to end. Meeker's grand plan to mark the trail with iron pipes fell apart almost before it began. The plan to write yet another book failed, and finances were a constant struggle. William Mardon got ill, and word from Washington, DC, was disappointing. Despite the dismal litany, there were some crowning successes. In 1910 Meeker made the first modern effort to map the Oregon Trail in its entirety. His end product was a map some two hundred feet long that he hoped to roll out on the floor of the House of Representatives as the ultimate visual aid. This map survives today.[1] He came up with a cost estimate for the seven hundred monuments he felt would be needed to properly mark the trail. In 1911–12 he ventured into the motion picture business, making a movie of the outfit and the trail. Most of all he kept the idea of a public obligation to save the Oregon Trail before the nation with his unprecedented publicity campaign.

Planning the Second Expedition

In a January 24, 1910, letter to Harvey W. Scott, the editor of the *Oregonian*, Meeker spelled out the impetus for a second expedition. He wrote that there was a bill pending in Congress appropriating $50,000 to mark the Oregon Trail, but that it contained a provision stating that the appropriation could not be spent until the Secretary of War was satisfied the project could be completed without any further funding from Congress. Meeker wrote it would require a second trip over the trail to determine exactly where monuments or markers would be required, how much they would cost, and, if costs exceeded the appropriation, to make an effort to obtain the difference from private donations. He was hopeful the bill would pass, and felt that gathering this information would disarm the criticism some congressmen made that the appropriation "was only an entering wedge for a larger sum." The plan, he told Scott, was to leave The Dalles, Oregon, around March 10 with his team and two assistants and again drive over the old Oregon Trail.[2] There was no mention in this letter of another equally pressing goal—paying off his debts from the AYPE financial fiasco. Embarrassment or subterfuge, or both, may have accounted for the omission.

Meeker's ambitions were, as usual, quite grand, but his wallet was empty. He sent a circular letter to businessmen and friends throughout the northwest outlining the proposed trip and soliciting contributions to the cause. Meeker tried to entice them with the promise that he would publish the names of the donors in his new book. Old friends Henry Hewitt and Robert L. McCormack contributed $200 each and

Dr. Nelson Blalock promised $500 from the Walla Walla Commercial Club. Several others contributed smaller amounts. Ezra offered Alden Blethen, publisher of the *Seattle Daily Times*, exclusive rights to weekly installments of this new book, with illustrations, if Blethen would sponsor the expedition to the tune of $2,500. When Blethen declined, a similar offer was made to Harvey Scott of the *Oregonian* and to the editors of the Kansas City, Missouri, newspapers. None accepted the offer.

Meeker enlisted his granddaughter, Olive Osborne, to type the manuscript of the new book that he planned to write day to day as he went along. Ezra wrote his daughter,

> I feel it in my bones that I can write a more salable work than any I have before written; of course it could not be ready for publication until I reached the Missouri River and not until I had retraced the Trail and marked it.... The plan of my new book seems to have taken possession of my mind like experience with the writing of the Tragedy...with this difference that I enter with more confidence the writing of "The Pioneer Way" than any other work I have written and think if I want a publisher, one can be had.[3]

A short time later the title had changed. It was now to be called "The Trail That Led to Empire." He told Blethen, "I sold 19,000 copies of my book, 'The Ox Team'; I feel it in my bones I will sell twice 19,000 copies of 'The Trail.'"[4] The letters make clear that Ezra sent Olive Osborne regular installments to transcribe at least through Oregon. In early May, however, Olive wrote her grandfather telling him that she was leaving Seattle for the east coast and Europe and would be not be available for more work. When Meeker reached Idaho the talk of his new book began to disappear from his letters. The fatal blow to the project came when his assistant, William Mardon, landed in the hospital, forcing Ezra to take on all the camp and expedition duties. Meeker was alone for nearly two months that summer. With no one to share the workload, he had little time for writing, and the work he hoped to publish on his eightieth birthday faded away.

Meeker initially wanted three teams to accompany the expedition, two for camp equipment and a light rig for his personal convenience and comfort. He intended to carry a compass and surveyors chain and to actually survey parts of the trail. Surveying was a skill Meeker had learned in Iowa in the winter of 1851–52 and that he took up again in the 1860s in Washington Territory where one of his contracts had him surveying the site of future city of Tacoma. David B. Miller, Division of Survey, General Land Office, Washington, DC, was contacted regarding the plan to map the trail in detail and "how best to make the report so to receive the approval of the Sec. of War."[5]

The original thought was to place wooden posts along the trail as temporary markers, similar to those used in a number of sites in 1906. Meeker planned to put one at each intersection where a current road crossed the trail and one at each site of historical importance. Discussions with Hiram Chittenden, Seattle District engineer for the U.S. Corps of Engineers, and City Engineer Reginald H. Thompson, suggesting that iron would be less liable to vandalism and not much more expensive, steered him toward the idea of using hollow iron pipes.[6] Various sketches of possible markers were drawn up by the Seattle City Engineer's Office for Meeker's approval.[7] He chose a simple hollow galvanized iron pipe with an inscribed brass cap on the top. An order was placed with Frink Iron Works in Seattle but bids were also solicited from the Willamette Iron Works in Portland and the Walla Walla Iron Works.[8] It is unclear who actually constructed the markers, but at least ten were produced and placed, all in Oregon.

George Himes, Secretary of the Oregon Historical Society and a friend and ally in the trail cause, was invited to accompany the expedition through Oregon, suggesting "a run out on the Trail for a couple of months where he could use his legs and run off some . . . surplus fat."[9] Ezra was a man in a hurry, telling folk, "I am now in my 80th year and can't wait."[10] He returned to Seattle from southern California at the end of February 1910 to wrap up loose ends. The ox team and wagon were left in the care of William and Cora Mardon to ship from California to The Dalles, Oregon. Preparatory instructions were sent to Mardon:

> I hear the weather is not broken over the mountains and if you find it uncomfortable to camp, then better take a room and put the cattle in the stable. Get what repairs are needed to the wagon done and arrange for prompt shoeing of the oxen; Keep an eye out for a mule team as I expect to be in position to buy one; sent $10.00 to Portland for Mrs. Mardon as requested.[11]

Unspoken in the flurry of preparations and high hopes was the fact that the large debts from the Alaska-Yukon-Pacific Exposition remained unpaid.

March 1910: The Expedition Begins

Meeker stayed in Seattle just long enough to vote in the local elections. He left March 12 for The Dalles, Oregon, where he reunited with the Mardons. Three days later he wrote George Himes, who had apparently decided to stay in Portland and not walk off that excess fat, with a description of his somewhat frantic activities at The Dalles. He gave an hour-long talk to thirty-one ladies of the Old Fort Dalles Historical Society and received ten dollars from the society (all they had in the treasury) and fifteen dollars from Mrs. Lord. The next day he journeyed five miles upriver to the old Native American village of Wishram, made famous by Washington Irving in his book *Astoria*. The visit to Wishram distressed Meeker so much that he wrote to Samuel Hill, a railroad executive and the president of Washington State Good Roads Association. (Hill paved the first roads in the state in 1909—ten miles worth at his Maryhill estate. Wishram was nearby.) Meeker was unhappy that the railroad siding at the site of the old village was now called "Spedis," and he asked Hill to request the railroad people restore the original name. He succeeded. Today the little town and railroad siding is called Wishram. Meeker also visited Pulpit Rock, which he called the Jason Lee rock pulpit, and had a photograph taken before going on to tour Fort Dalles. This volcanic rock is on a hillside overlooking the The Dalles' downtown. It can be easily climbed and supposedly the missionary Jason Lee gave sermons from the rock. His brother Daniel Lee, who was the resident missionary at The Dalles, also likely did so.

One final duty remained before heading east. On March 19 Meeker placed his first iron pipe marker at "Harbor Rock" on The Dalles waterfront. He was aware that the rock was submerged under some ten or twelve feet of high water at spring runoff, but it was appropriate for a marker for several reasons. The location was significant because Harbor Rock was used as a survey reference point when The Dalles was platted, and it also marked the entrance to the steamboat harbor at The Dalles. But to Meeker its main importance was that, "Here at this point the immigrants embarked on scows or rafts to float down the Columbia to the Cascades," just as he had done in 1852. Meeker noted in his journal, "Busy day: set 3 foot galvanized 1 ½ pipe as first marker in crevice of Harbor Rock. We needed to chisel but little to let the pipe in the crevice in an upright position after which filled the surrounding space with cement or rather concrete."[12] The area has since been completely filled in. The rock, if it still exists, is now under the Post Office parking lot located at 1st and Union. According to a historian at the Fort Dalles Historical Society, the pipe and cap disappeared within a week of being placed.

On March 21 Meeker broke camp at The Dalles and drove ten miles to Fairbanks. On the way out of town he found the boulder that he and Mardon had painted four years earlier. The paint was nearly worn off so they stopped, unyoked the team, put Mrs. Mardon to cooking dinner, and set "out with the cold chisels and dug the letters into the rock 'Oregon Trail,' so they will be legible for a century."[13] That night they camped at the identical spot where Meeker had stayed in 1906. The place was then called "Coopers" or 10-Mile Creek, where he put up his second galvanized iron pipe, "set in the ground 2 feet tamped with earth to within six inches of the ground and remainder with cement."[14] A third marker was placed near Miller's house opposite the station of the Oregon Railroad and Navigation Company.[15]

Wasco, Echo, and Walla Walla

Meeker continued east placing markers at Biggs and on the left bank of the John Day River. At this early point in the journey the ox Dave went lame, forcing the hiring of a horse team for a few days. Mardon took the hired team, oxen, and wagon to Well Spring, located the boulder they had painted four years earlier, and chiseled the letters into the stone.[16] Ezra went ahead by train to Echo where he placed a marker and lectured at the local school for forty minutes. The hired horses were sent back to their owner and replaced with a mule team, funded by his son-in-law, Eben Osborne. The mule team traveled about seventy miles, as far as Walla Walla, where they were sold due to financial difficulties.[17]

The next stop was Walla Walla where the outfit camped opposite the post office, intending to stay for at least two weeks. Meeker's older brother, John Valentine, was living there at the Odd Fellows home, and Ezra anticipated a good visit with him.

There was also a tantalizing offer of $500 of financial aid from the Walla Walla Commercial Club holding him in the city. Unfortunately, it failed to materialize, forcing the sale of the recently purchased mules. Although Meeker made forty dollars on the transaction, he regretted the eighty-mile, ten-day detour he had taken. The lack of financial

Well Spring boulder. *Ventures and Adventures*

aid forced another difficult decision. He could no longer afford two helpers, and the wagonload had to be drastically lightened if the ailing ox Dave was going to be useful, so Mrs. Mardon was sent home to New York. Meeker noted in his journal that by doing this he had reduced the wagonload by about four hundred pounds or nearly one third. As it is certain Mrs. Mardon weighed nowhere near four hundred pounds, other unnamed items had to be jettisoned to achieve that loss. After some discussion her husband decided to remain. Meeker wrote that the separation was arranged "good naturedly."

Pendleton to La Grande

On April 14, 1910, the teachers at Grove School in the small village of Milton-Freewater told their students that a pioneer of 1852 was coming with a covered wagon and an ox team to tell them about the Oregon Trail. The children marched outside and sat down in rows on the grass. Just up the street they could see the pioneer digging a hole in the earth and putting a long pipe in the hole. He then packed the dirt around the pipe, cleaned his shovel, climbed aboard the wagon and drove down to where the children were seated. The superintendent greeted him as the children clapped and cheered. Ezra entertained them with stories of pioneers and the trail and explained what he was doing. He concluded, as he often did, by having the students each place a small rock at the base of the pipe monument, building up a mound of stones. He felt this gave the children a sense of ownership and personal responsibility for preserving the memory of the trail into the future. The program concluded with the singing of *America*. Meeker camped in a grove of cottonwoods near the Little Walla Walla River and lectured that night in a hall and again in a local church.

Years later Effie Ritchey, one of the children present that day in 1910, pointed out the site of the monument at 139 South Main Street. In 1989 while doing landscaping at that location workers uncovered an old pipe, perhaps Meeker's "monument." Today a granite stone stands there, and inscribed on it, along with the text, is a picture of the "pipe monument" surrounded by a mound of stones.[18]

In 2016 the Oregon Trail marker at Toledo, Washington, was refurbished under the direction of the Daughters of the American Revolution (DAR) and rededicated. At that ceremony the local school children placed small rocks at the base of the monument, unknowing that they were replicating a tradition started by Meeker over a century ago.

Meeker next arrived for a three-day stay in Pendleton after stops at Weston, Athena, and Adams where he placed pipe markers.[19] Placed opposite the railroad depot, the Adams marker was the last iron marker placed by the expedition. Apparently, none of these iron markers survive today.[20] From Pendleton, Meeker wrote his daughter saying he was well and that although Mardon's cooking wasn't as elaborate as his wife's it was good enough.[21]

April 23 found Ezra and Mardon nooning in the Blue Mountains where he wrote Senator Piles about the monument bill currently under consideration in the Senate and pointed out to the Senator that "I do candidly believe my digestion would be better could I but hear that the bill had passed the Senate."[22] At Meacham they found the monument placed there four years earlier defaced by someone carving a name into the stone. "[W]ith cold chisel and hammer soon removed it after which photographed the monument and moved across the RR & Meacham creek to the reputed 'Lees Encampment,' and lunched."[23]

Meeker spent three days in La Grande camped next door to the post office. He had $300 cash on hand and between sales and lecturing was earning a little more than expenses, so he gratefully accepted an invitation to spend the nights of April 29 and 30 at the Hot Lake Sanitarium, just up the road, courtesy of Doctor William T. Phy, the medical superintendent and manager. Ezra and Mardon were given free rooms and free access to the dining room. The two men had their picture taken with the outfit in front of the building in the afternoon.

That evening Ezra gave a lecture to some one hundred guests, sold books and post cards, and put one hundred dollars in the bank.

However, serious trouble was just up the road. The outfit departed at six o'clock in the morning, but just a few miles along, at the mouth of Ladd's Canyon, Mardon announced he badly needed to consult with Dr. Phy and that he was certain he would need an operation.[24] Meeker assented, gave Mardon fifty dollars and drove up the canyon. Mardon was indeed operated on the next day. The nature of his affliction was never stated. It would be three weeks before he was able to rejoin Meeker, but then he was so weak he was of little use. In the meantime Meeker was alone with his dog Jim, the wagon, and the oxen.

Baker City

Two days later, in a rainstorm, Ezra drove the team into Baker City. He had run out of copies of *Ventures and Adventures* and decided to wait for a new supply to arrive from Seattle and to see how things went with Mardon. He put the team in a stable and checked into the Packwood Hotel. With $225 on hand, he felt that he could wait a few days. But the books, and thus an income from their sale, were slow in coming. He waited six days, then wrote to his grandson, Joe Templeton, asking him to check on the book situation. He also sent a scathing letter to Mr. Calder, his Seattle publisher, telling him that without books to sell, he could not hope to pay off his debt. He reminded Calder that he had promised to have five hundred copies bound

Books in the wagon on the street in Baker City. *Author's collection*

and held for prompt shipment when requested and asked Calder to increase the number to be bound to one thousand. A box of fifty copies arrived the next day. Meeker used the enforced delay to work on the blueprints of the government survey for seventeen townships showing the location of the trail in Oregon, the first installment of the grand map he intended to roll out on the floor of the United States Congress.

Mardon's recovery was slow. He wrote Meeker, "I have been able to sit up and I am pretty weak." Three days later the news was less encouraging. "I am not getting better as fast as I expected; the cut is healed up but the wound is still badly swollen. I put my feet on the floor today for a short while. I am feeling well and eating regular now. Only that the wound is still tender and a few more days rest here will count at the present time." Nearly a week later he reported, "I am feeling all right except where I were operated on is still tender and a little painful."[25]

While in the hospital, Mardon asked Meeker to send his wife twenty dollars.[26] He received a stern "no."

> I cannot send $20.00 to your wife… speaking plainly, Mrs. Mardon is at home and among friends, and strong and able to take care of herself and knows the arrangement between us and she ought not call for money nor now expect it when she knows of your experience in the hospital.[27]

Meeker was stuck. Lack of finances had ended the plan to mark the trail, so he turned his attention to locating and mapping the trail on blueprints of township plats and marking the road crossings from information obtained from county surveyors. He hoped to time his journey to be in Cheyenne, Wyoming, on August 8 when President Roosevelt was scheduled to speak there, and in Osawatomie, Kansas, on August 30, at the dedication of the John Brown Battlefield, where Roosevelt was again scheduled to speak. Meeker envisioned selling a thousand copies of *Ventures and Adventures* at these two events.[28]

On May 6 he addressed an audience of three hundred in the Baker City library auditorium. Two days later he attended a Presbyterian service. At the close of the sermon the minister recognized Meeker in the audience, introduced him to the congregation and invited him to speak at the conclusion of the evening service. Meeker wrote, "A curious incident occurred just as the choir were closing, the minister asked me if I was a church member; I told him I was a Unitarian and in his introductory remarks mentioned my being a Unitarian. This is some different than what it would have been years ago, a Unitarian occupying a Presbyterian sermon—the world does move."[29] The next day he wrote Carrie to tell her how well his talk at the church was received. A few days later he again wrote his daughter telling her that he had been delayed ten days, but did not regret it as he might not have delivered his "sermon" which had made quite a "stir" in Baker City.[30]

Another box of books finally arrived and Meeker spent a week selling on the streets. Ezra wrote Mardon that if he was well enough by May 17 to drive the team and wagon alone to Durkee, he would go ahead by rail and make advance arrangements. If not, Ezra advised him to remain in the hospital and said that he would drive the team alone, and that he had booked a program at Durkee for May 18. He also sent along five dollars.

Mardon was not well enough to leave the hospital, and Meeker wrote him to stay until he recovered. The route ahead promised to be challenging after floods had taken out the bridges over the Burnt River, necessitating a long detour. Ezra decided to ship by rail to Parma, Idaho, about fifty miles from Boise. He rode in the car with the wagon and oxen, munching on a lunch and a bottle of buttermilk given him by his landlady. Volunteers loaded the wagon and oxen onto the train. Despite his difficulties, Meeker left Baker City a happy man. His map work was going well and he had determined how many monuments would be needed in Oregon. At Boise he hoped to obtain Idaho maps and continue that part of his work. However, he was still steaming about Calder's failure to deliver the needed books in a timely manner. He complained that this failure cost him one hundred dollars.[31] He wrote Mardon of his new plan to ship the outfit by rail over the Rocky Mountains from Pocatello to Cheyenne but swore him to secrecy, even from his wife, as he advertised the expedition as a drive over the Oregon Trail, not a train ride. He assured Mardon his job was still secure, but not until he was completely healthy.[32]

The train departed on May 16. Meeker wrote in his journal, "Looking from the car door as we come down Burnt river and through the mountains I realized more than ever how exceedingly rough it is and again how on earth we did manage to get through in early days."[33] It was a 104-mile train trip to Parma. He hired two men to unload the wagon and team and drove to a nearby restaurant and began selling immediately. He was so busy he did not even have time to inspect the monument he had purchased and sent to Parma in 1906. He slept that night in the wagon. In the morning he hitched up the team at 3:30 a.m. and drove nine miles to Notus. Meeker commented that he had a magnificent view of the Halley comet "tail that reaches full 90° and seemed to touch and almost blend in with the Milky Way."[34] He rested in Notus for seven hours and continued on to Caldwell.

Mardon Returns

On May 19 Mardon arrived by rail and rejoined the expedition. The tent was set up and Meeker began to "feel at home again." The next morning Ezra went into Boise to get the mail, check for a freight delivery of books, and enquire about purchasing township maps. Governor Brady arranged for a photograph to be taken of the two of them and invited Ezra to dinner, after which he returned to Caldwell. The next few days were spent selling books and securing township maps from the Idaho Surveyor General showing all the old roads, thus enabling Meeker to calculate how many monuments would be needed in that state. They were welcomed heartily in nearby Nampa where a local farmer invited them to breakfast and the mayor said, "You can have anything in the town," and invited Ezra to speak at the high school. When the expedition arrived in Meridian Ezra took a trolley car to Boise and arranged for a camping place. That night Meeker lectured in what he called "the hall of a moving picture show." He took his cot into the hall and slept there, leaving the wagon to Mardon. It is virtually certain, as Ezra slept that night in the theater, that in his wildest dreams he would not have imagined himself in the motion picture business within two years. But that is what fate had in store for him. Ezra wrote that they "overslept" but were still able to start for Boise at 5 a.m. Mardon slept in the wagon most of the way. At the end of the day, the travelers set up a "splendid camp" in the shade of a box elder tree at the corner of 9th and Bannock.

The dispute with Calder escalated at this point, and the hard edge of Meeker's personality emerged. There was much correspondence between Meeker and Joe Templeton about needed supplies after Ezra departed from Baker City, but the overriding topic was Mr. Calder's inability to get books shipped as needed. Meeker told Joe to threaten Calder with a lawsuit saying, "This thing of delay has been tolerated too far already and has damaged me a hundred dollars already and I am in no mood to dally further."[35] Calder was fortunate that he was only threatened with a lawsuit. It was a venue in which Meeker had much practice. In Washington territorial days alone Meeker went to court fifty-three times. It was not an idle threat.

Mixed in with this verbal war with Calder and the Rainier Printing Co. are several letters and telegrams to Malcolm A. Moody of The Dalles, Oregon, suggesting that Meeker had arranged a way out of this jam if needed—a backup plan.[36] He was not totally out of books after all, as he had stored about a dozen boxes with Moody, who shipped them up the line as Meeker ordered them. The first order to ship a box forward came when Meeker was in Pendleton. Then there was a pause. When Ezra arrived in Idaho, and the problems with Calder continued, the orders resumed. On May 25 Moody billed Ezra $7.50 for storing the books, and then said, "Amount of above to be applied as subscription to cause of Oregon Trail." In short, there was no charge. As Meeker made his way east he left deposits of books and postcards at various storage houses. If a disaster struck and wiped out his wagon stock, he could replenish relatively easily. It was a technique that served him well over the next couple of years. Still, he continued to vent about Calder.

Templeton finally got Calder to put two printers to work, each preparing five hundred books, and Templeton informed Ezra that there were about 2,500 books left to prepare. The delay in getting inventory to sell prevented Meeker from paying off the Calder bill before he left Boise. The towns ahead were small, and he needed to build up a reserve to pay the costs of shipping the outfit by railroad over the Rocky Mountains to Cheyenne. Meeker believed he would sell out the entire edition well before reaching Kansas City. At the same time, work on his new book had ground to a halt as he found himself doing chores that the still-ailing Mardon could not do.

A good portion of Meeker's stay in Boise was devoted to the mapping project and a trip out to Sinker Creek, east of Murphy, to locate the trail in that area and visit the Otter Massacre sites.[37] He found he could get eight-foot tall granite markers inscribed for about fourteen dollars each from a local company, thus supplying needed information about cost for his report to the Secretary of War. Letters went back and forth to Congressman Humphrey discussing the prospects of getting the bill through the House of Representatives. On June 1 at four-twenty in the morning Ezra and Mardon drove out of town heading for Mountain Home where they spent the next three days.

Glenn's Ferry to Pocatello

From Glenn's Ferry Meeker reported sales were better than expected. He had a picture taken of himself on top of the hill on the south side of the Snake River looking down at Three Island Crossing. This picture was later turned into a postcard. He also received a letter from the Hot Lake Sanitarium containing "two photographs of the 'Meeker outfit' and the Sanitarium buildings," one for himself and the other for Mardon.

During the next stop in Twin Falls Meeker spoke to a crowd at a Flag Day celebration at the foot of Shoshone Falls. He sent a one hundred dollar money order to Joe Templeton with instructions to pay off Calder's bill and promising that Mrs. Ackerson was next in line. He then jumped ahead to Pocatello to pick up a book shipment, leaving the team eighty miles back with a slow-moving Mardon to follow. At this point he had enough funds on hand to pay his way to Cheyenne, but needed sales in and around Pocatello to replenish his dwindling cash supply.

On the last day of June, Meeker and Mrs. Drew W. Standrod, president of the Ladies Club of Pocatello, rode by automobile north of town to search for the old Fort Hall site. Meeker was of the opinion that this was the single most important historical site on the entire length of the Oregon Trail, and it needed to be marked. Built as a fur trading post by Nathaniel Wyeth in 1834, Fort Hall became a major resupply point in the 1840s and 1850s for emigrants on the Oregon-California Trail, and the departure point for wagons heading south to California. An estimated 270,000 emigrants stopped there. The problem was no one knew for certain where the old fort stood. There were no visible ruins and there was much conjecture as to its location. The Ladies Club had erected an Oregon Trail monument in Pocatello in 1909, originally on the high school grounds. (It has since made a couple of moves and today stands inside the Fort Hall replica.) They placed a second monument at what seemed to be a promising location but was not the Fort Hall site. Meeker confirmed the error but had his picture taken next to the monument anyway. He urged Mrs. Standrod to keep looking.[38] Six years later he would return to Pocatello and finally succeed in locating the original site. The 4th of July found the team in the Pocatello city parade, with Meeker speaking at the ceremonies following the governor.

Cheyenne, Wyoming, and Greeley, Colorado

The expedition arrived in Cheyenne about a month too early to coincide with Theodore Roosevelt's August visit. The outfit traveled there by train, avoiding a month of hard travel over the mountains that would have severely taxed both Mardon and Meeker. The cost was $120. Camp was set up on a vacant lot opposite the post office, book sales were good, and for once there was an ample supply. Here Mardon decided to recuperate in New York where his wife was also ailing. He drew his salary of sixty-eight dollars, leaving Ezra with twenty-five dollars on hand but confident that he would not go broke.

Mardon was sent off with a letter of recommendation that read in part, "After four years camp life together one naturally gets used to each other's foibles…I have found you to be honest and capable and devoted to the interest of the work in hand and as I have said am sorry to see you go."[39] Mardon replied from New York, thanking Ezra, saying his wife had recovered from her illness, and that the doctor advised him to take a month's rest. He apologized for being surly or sharp at times but assured Meeker "that I were as quick to resent any criticism or injury to you or your laudable work and effort to mark the old trail and trust and hope your efforts will finally be crowned with success."[40]

The stay in Cheyenne was a repeat of Boise. Meeker lectured at the library with the current governor and two ex-governors in attendance. He addressed the local DAR chapter, and obtained township maps from the Surveyor General's office that covered the route of the old trail in Wyoming. He had his picture taken with the governor in front of the state house building and turned that photograph into a postcard.

On July 14 Ezra began a solitary sixty-seven-mile drive to Greeley, Colorado, stopping at small towns along the way to sell and lecture. He slept in the wagon with his dog Jim for company. He arrived five days later and set up camp near the courthouse. Greeley was founded by Nathan Meeker, a distant relative. It was also the city to which his nephew, Frank Meeker, initially fled in an attempt to escape the attendant notoriety after being shot on the streets of Tacoma by his mistress.[41] Ezra had taken Frank and his widowed mother into his home after the death of his brother in 1860, helped raise him, and paid his way through Cornell University. Frank had subsequently moved to southern Oregon, but Nathan's daughter Rozene paid Ezra a visit and they dined together.

While in Greeley, Meeker decided to ship the outfit by rail to Kearney, Nebraska. He concluded not to employ any more help, saving around one hundred dollars a month in expenses, but it meant all his meals would be at restaurants. Shipping to Kearney left him with enough spare time in his schedule to allow a leisurely drive through Nebraska. He told Carrie, "I left Cheyenne way broke after paying off Mardon and paying for the township tracings for Wyoming but am again in funds… don't worry because I am alone for I will be in a thickly settled country and most of the time entertained by friends of the work."[42] Don't worry, indeed. Her father was seventy-nine years old, half a continent away, alone with an ox-team and covered wagon to care for, and struggling financially. It was not a situation to inspire confidence.

Nebraska

Meeker spent the month of August in Nebraska, making extended stops in Kearney and Hastings. He was reluctant to hire a crew because his expenses were so much lower without, and the townsfolk encountered were exceedingly welcoming and generous. Ezra noted at Hastings he hadn't purchased a meal in a week and had paid very little for feed for the cattle. Sales also were going so well that he invested $200 in printing 72,000 postcards. The work was done by Adam Breede, editor of the *Hastings Tribune*. Meeker considered this a good price, since he expected to sell them for $800 retail. Unfortunately there was a problem with the quality of some of the printing, about which Ezra complained, but Breede agreed the fault was his and refunded the thirty-four dollars in dispute.

More good news arrived: Joe Templeton informed Meeker that Calder delivered the last eight hundred books and that the account was paid in full along with some incidental charges. "The books are to remain in the printing office and are to be shipped upon your order, without further costs to you."[43]

On August 18 Ezra attended the Woodsman Picnic in Hebron, Nebraska, and the day was a smashing success. He sold four thousand postcards and over fifty books. He wrote Joe Templeton suggesting that perhaps he should use all his available funds to pay off Mrs. Ackerson's note, as she seemed to be impatient. He would trust to the future for continued solvency, as sales of postcards alone were meeting his expenses.

The lucrative day at the Hebron picnic caused a slight change in plans. Instead of continuing down the trail, he detoured north toward Lincoln to test the financial waters in that venue. On August 22 at Strang, Nebraska, he engaged an automobile to drive him around and a railroad car to move the team and wagon to the various towns he planned to visit. This resulted in $125 in expenses. He wrote Carrie, "Traveling this way would make it much easier for Dave and me, to say nothing about Jim who will persist in running his legs off after anything in motion."[44] Each stop gave him a site for his wagon and oxen, a place to set up his teepee, the privilege to sell his book and postcards, and allowed him to "preach Oregon Trail" to the crowds. He also volunteered to join any parade that would have him.

While in Lincoln, Meeker obtained a set of township maps for Nebraska and, when time permitted, worked out the costs of marking the trail in that state. To gather information, he exhibited his maps at the various state and county fairs he attended, displaying them on a twenty-foot board. He would then compare notes with the "Old Timers" who came in droves to see the outfit and discuss the trail. He also took time out to do field work on his own, driving out to personally check various important sites and locations.

On September 6 Meeker noted in his journal that he had finished his map work for Wyoming and Nebraska. Then George W. Martin, secretary of the Kansas Historical Society, sent him a detailed map of the trail in Kansas, essentially completing the entire mapping project. It earned Meeker a fair share of national publicity. Mardon read about it in Stamford, Connecticut, where he and Cora were staying with Cora's brother. Mardon wrote, "it were quite a long article well written and giving particulars of what you had done and what you were striving to do. One thing I noticed in particular were that you had a map 370 feet long showing the Trail in detail to lay before Congress (well I reckon that is some map, eh)."[45]

Kansas to Missouri

Meeker wrote very little about his trip through Kansas beyond the fact that he was at Topeka from September 11 to September 17 doing a stint at the Kansas State Fair. Apparently, he failed to attend the John

Brown Celebration in Osawatomie, and once again missed seeing President Roosevelt. George Martin of the Kansas Historical Society wrote a letter of recommendation for Meeker while he was in Topeka, revealing some insight into how Meeker and his crusade were perceived by friends of the cause. Meeker had applied too late to reserve a space at the Illinois State Fair in Springfield. Martin wrote to Jesse Palmer, secretary of the Illinois Historical Society, to ask if she could help him.

> The old man passes for a freak, but all the same he is making a great mark historically and otherwise. On his return from Washington two years ago, making the distance from that city to Topeka with his oxen, he stopped here four or five days. We found a young man with an automobile to drive him about town. In one day he visited seven of our public schools and at each place made a twenty-minute speech on the saving of that country to the United States. The work is purely and solely educational and patriotic. Can you help him get onto your fair grounds? Help the old fellow along, if you can. There is no fake about him—just a straight, patriotic old crank.[46]

The expedition arrived in Kansas City, Missouri, on September 23. Meeker was led to believe the *Kansas City Star* would take interest in the work and publish an article worthy of his mission. He drove the wagon to their office and had a picture taken, but was disappointed in the article. The city government had changed since his visit in 1908 and his effort to get a permit from the mayor met with a rebuff. Meeker eventually got his permit by going to the city council and bypassing the mayor, a technique that he employed in New York City in 1907 and which he would find useful at various stops down the road.

Mardon Returns, Take 2

Mardon wrote Ezra on September 27 from Connecticut, "Well Mr. Meeker … if you have the old position open for me as you said you would when I left Cheyenne, Wyo. I would like to go back to work again as I have recovered from the operation and have had a good rest."[47] Ezra immediately sent him a telegram. "Position open, at forty dollars per month. Answer quick. Important parade Wednesday. Draw for thirty dollars if needed."[48] Ezra wired Mardon forty dollars for train fare and William arrived on October 5 just in time to lead the ox team in the Industrial Parade. He came without Cora, who remained back east. While in Kansas City Meeker was invited by

Mrs. Elizabeth Butler Gentry to attend a DAR meeting about the Santa Fe Trail. She was a staunch friend of the trail whose aid proved to be invaluable two years later.

The question of what to do next and how to pay for it always loomed large in Meeker's letters. Even though the first printing of *Ventures and Adventures* was now fully paid, the vexing Ackerson loan and other outstanding debts were still due. Meeker was sending a steady stream of money orders to Seattle. He was also nearing the end of his supply of books and turned to M. A. Donahue & Company of Chicago, the company that printed the second and third edition of *The Ox Team* for him in 1906 and 1907. In January, when he was still in California, Meeker had sent Donahue a copy of *Ventures and Adventures* and asked for a bid on printing five thousand copies. Donahue quoted Ezra a price of twenty-eight cents apiece for the first five thousand and then seventeen cents afterwards. This was a price he felt he could afford. He had been selling the book at a dollar a copy and they were moving slowly. If he got the price down to seventy-five cents or lower he felt he could sell them by the thousands. The cost of the composition and plates and the printing of the first half of the edition would total $1,300, to be paid when the contract was signed. To his mind, it was just a matter of getting these up-front funds. When it came time to leave Kansas City Ezra expressed disappointment in his book sales to both Joe and Carrie. He reminded them, though, that four years earlier, of the 19,000 copies of *The Ox Team* sold, 16,000 had been sold east of the Mississippi River.

ILLINOIS TO INDIANA

From Kansas City Meeker sent Mardon and the team on to Independence, where he met up with them. By October 27 they were at Carrollton, a small town about sixty-five miles northeast of Independence. If sales warranted, Meeker intended to continue the drive all the way to Springfield, Illinois, but a cold snap abruptly changed the plan. It laid down a quarter of an inch of ice on the dirt roads, so instead the outfit went by rail to Springfield, 225 miles to the east. The cold weather also sent the seventy-nine-year-old Meeker to a hotel. Mardon got to sleep in the wagon. Ezra mused that his assistant could get along all right with the doubled up bedding and a coal oil stove. The cost of this 225-mile jump was fifty dollars. The hope was to stay a week or more in Springfield and then start working toward Cincinnati. All four of the city newspapers gave him glowing coverage, resulting in an uptick in the sale of his books. On November 2 he spoke to a small but appreciative audience at the YMCA hall. This speech was printed in full in the *Illinois State Register* on November 6, 1910. The next day he addressed eight hundred high school students.

Up to this point in his travels, Meeker had contracted with a dozen different companies and shops all across the country to print his postcards. Many times he was extremely disappointed in their quality, once to the point of suing for $600 in damages. While in Springfield Meeker met Murray Bloom, a salesman for Curt Teich & Company of Chicago, discussed business, and signed a contract for ten thousand sets of colored post cards. Going forward, the company printed all of Meeker's postcards and he raved about their quality.

On to Cincinnati

After four days in Terre Haute and five in Indianapolis, the expedition moved on to Cincinnati, where the city fathers welcomed them with the ever-elusive permit. Meeker and Mardon hit the streets in a snowstorm. In three hours they sold forty-seven dollars in postcards and five dollars in books. This success prompted a decision to stay a month. Ezra rented a warm room for himself at the Dennison Hotel and a nearby one for Mr. and Mrs. Mardon for ten dollars a week combined. Ezra's cook was back. Mrs. Mardon had rejoined the expedition, and Ezra paid her five dollars a week for meals.

As the hearing before Congress loomed closer, Meeker wrote the Twin City Granite Works in St. Paul, Minnesota, and sent them a copy of the bill in front of Congress. To adequately mark the trail, he estimated that it would take seven hundred granite markers, set in concrete and inscribed with sunken letters "Oregon Trail." He requested a cost estimate of stones of various sizes and supplied them with a list of cities along the trail to which the markers would be shipped.

Ezra also contacted Edward Jay Allen, his old 1852 Oregon Trail partner, in Pittsburgh and suggested a ten-day visit around the time of his eightieth birthday. He then fired off letters to Congressman Humphrey and Senator Piles outlining his progress to date and requesting a hearing before the appropriate committees. He informed them that he would be eighty years old on December 28 and time was running out. When a local theater offered to pay him ten dollars a night for illustrated lectures, he decided to work lecturing into his schedule as time permitted. There were eighty-three theaters in the city.

A series of disappointments followed in quick succession. Edward Jay Allen could not join him for a birthday celebration although he urged Meeker to stop in Pittsburgh on his way east. He received word that his older brother John Valentine Meeker was seriously ill and not expected to survive. Next came the refusal of Joe and Carrie to send him money to print a new edition of *Ventures and Adventures*. And the final blow was a letter from Senator Piles on December 31 telling him not to come to Washington, DC, as there was no hope of favorable action on the bill during the current session of Congress. It was not a happy eightieth birthday for Meeker.

CHAPTER 7

The Second Expedition, 1911

The New Year started with a disheartening report from Congressman Humphrey to accompany the one Meeker had already received from Senator Piles. "I doubt...if I shall be able to secure a hearing for you.... I can do this, however: if you desire to have a statement submitted and will place it in writing, I can put it in the Congressional Record."[1] Meeker acknowledged chances for passage of the bill this session were now slim and informed Humphrey that he would submit the statement. For over a year Meeker had been telling people far and wide that he was going to Washington City to testify in front of Congress and present the results of his work to the House and Senate committees. It had taken all of his resources just to reach Cincinnati. Now he was told not to come. The disappointment must have been deep.

Nevertheless, on January 13 Meeker wrote Humphrey describing his nearly two-hundred-foot-long map, his planned "campaign of education," and his expectation to spend the summer in the New England states.[2] Humphrey replied the next day saying he would insert Meeker's statement into the Congressional Record at the earliest opportunity and promised that if the bill did not pass he would introduce it at the next session.[3]

Meeker urged the congressman to reintroduce the bill and added prophetically,

> I will remain east and next summer try [to] stir them up in the New England states...and will do good emphasizing the idea of a national road "Pioneer Way" that I have harped on so much. I thoroughly believe the traveling car is soon to be a factor in transportation and the sooner the people begin to realize it the better. I hope you will give some thought to this greater monument "Pioneer Way" that will naturally follow the completed lesser work.[4]

Meanwhile, he needed to survive the winter in Cincinnati. As always, money was in short supply. Meeker was living in the Denison Hotel. William and Cora Mardon were in a nearby apartment where Ezra went each day for breakfast and dinner. Troubles with his books were unending. Three boxes were wandering around the country, lost on the railroad lines, until eventually reappearing with an unexpected bill for $20.16. When the boxes were opened, Ezra discovered that the bindery, in sealing them, had driven nails through several of the books, rendering them valueless. And the bad news kept coming. On January 5 Ezra learned that his brother John Meeker had died. While he had been warned of John's condition, the news still came as a shock to him.

On January 5 Meeker signed a contract with the Columbus Lecture Bureau and began a series of speaking engagements that served the dual purpose of earning money to help pay expenses and furthering his campaign to educate the public about the importance of the Oregon Trail. The topics of his lectures were "The Lost Trail" and "The Winning of the Farther West." He told the lecture bureau, "I have slides to illustrate where convenient but lecture with or without as occasion may require. If you can point the way, I am yours to command."[5] Mardon was given a note that authorized him to sign and arrange speaking contracts, and was sent to Hamilton, Ohio, Ezra's birthplace, to drum up business and secure a permit for a future visit. Meeker quickly filled his calendar through January 18 with engagements close to Cincinnati. These lectures were in motion picture theaters, and gradually it began to dawn on Meeker that this new medium might be a better way to reach the public with his mission.

Meeker and Mardon quickly learned the ropes of the lecturing business. They sent large poster-sized photographs of the ox team to the theaters. Written contracts, rather than verbal agreements, became the norm. Meeker signed some of these contracts; Mardon signed others. They had leaflets printed to advertise the lectures, and obtained letters of introduction from the Cincinnati Men's Business Club. Meeker found he could speak an hour without boring his audiences. He became adept at operating the stereopticon and showing his slides as he lectured. Usually the theater showed two motion pictures with Meeker's lecture in between. When he could find a vocalist to lead the audience in singing, Meeker projected the lyrics to the song "Fifty Years Ago" onto the screen with fifteen illustrated slides. He usually received 40 percent of the receipts and in this way kept adding to his funds.

In mid-January the expedition left the confines of their comfortable Cincinnati quarters and started working north through Ohio. (Cora Mardon remained in her apartment for a time.) The plan was to drive to Lake Erie, lecturing and selling as they went. Then they would ship by steamer, first to Detroit and then to Buffalo. From there they would drive across New York after which, Ezra wrote, "I expect to spend several months in the New England states, where there are so many senators to the square mile." He also had a potential support group in New England. Katherine Graham, who had served as caretaker for the Puyallup mansion and had attended to Eliza Jane Meeker's needs for several years, was living in Massachusetts. In addition, Ezra's granddaughter, Cora Osborne Brown, was living in Connecticut. New England was a section of the country he had never visited and he was looking forward to seeing it.

But first came Ohio. The expedition stayed in Dayton from January 23 to February 1 where the greeting wasn't exactly cordial. Ezra stayed in the Manhattan Hotel while Mardon slept in the wagon by the ox team. In response to an unspecified problem (though one suspects another permit dispute) Meeker complained to the local chief of police, "I don't believe you understand me or my work."[6] As they moved north Mardon drove the team while Meeker rode the train ahead and stayed in hotels. The expedition earned $150 over expenses for the month of January, which Meeker thought was pretty good for a winter campaign. He told his daughter, "William remains faithful and just as faithfully spends his wages, but now on his wife." In the first half of February the expedition netted $385.10 but Meeker lost Mardon again. "William has driven the team while I have traveled in the cars but last evening he received a telegram from his wife at Cincinnati to come forthwith and by telephoning ascertained she has been poisoned by canned oysters; he went at 3 o'clock this morning."[7] Two steps forward, one step back.

Meeker ran into Sunday Blue Laws for the first time in Springfield, Ohio, where he stayed February 22 through February 28. He wrote to an acquaintance, "Threats were made here by the law and order league to arrest me if I lectured here on Sunday. Those gentlemen found the same law would apply to the ministers of the churches and choir singers as would apply to me and the music in the theater and the thing was dropped."[8] He told Carrie the lecture went well. "I am a wonderful man, leastwise I am if the unanimous voice of the employees and owner of the Hippodrome Theater is true."[9] Meeker addressed eight hundred people over the course of the day at the Hippodrome, speaking a total of six hours in six separate appearances for which he earned thirty-two dollars. He told his daughter he was not particularly tired after doing so.

On the evening of February 28 Meeker took the train to Columbus to make advance arrangements. He left Mardon, who had rejoined the expedition after being assured his wife was in no danger, thirty miles back with the team. Meeker would stay here until March 24, lodging at the Metropole Hotel. He again ran into problems with city authorities, including more difficulties in securing a permit to sell his literature, and objections to giving Sunday lectures.

On March 13 he warned Carrie things in Columbus could potentially get ugly.

> I have here encountered a "fool" Mayor who forbids the sale of my literature on the streets and yesterday said he would run me into the lockup if I undertook to lecture here on Sunday. I had before had the question carefully looked into and knew he had no authority to stop me and so I told him that if he became a law breaker and interfered with me, I would "run him in through the courts," so don't be alarmed if you hear I have been arrested.[10]

Meeker decided to challenge Mayor George S. Marshall. He gave complimentary tickets to every member of the legislature, all state, county, and city employees, and members of the press—some four hundred tickets in all, in a theater that held 1,500 people. He further printed an open letter to the mayor in which he stated that he had leased the Board of Trade Auditorium for a lecture Sunday evening, March 19. And for good measure enclosed "complimentary tickets for yourself and lady."[11] George Harper, a city councilman who was running for mayor, sent Ezra a letter of encouragement. Ultimately Meeker gave the lecture to a small crowd and stayed out of jail. We assume the mayor did not attend.

A good portion of April was spent in Toledo. Meeker lectured twice each evening to what he estimated were audiences totaling four thousand weekly. At times he received such sustained applause that he

Dodger advertising Meeker's disputed lecture in Columbus, Ohio, March 19, 1911.
Author's collection

was embarrassed. Things were going well. Congressman Humphrey introduced a new bill asking for an appropriation of $100,000 and sent Meeker a copy of the February 25 Congressional Record which included four pages of Meeker's report.

When Meeker arrived by boat from Toledo, the *Detroit Free Press* reported on May 2, "With the snows of four score years drifted upon his head, Ezra Meeker, frontiersman, arrived in Detroit yesterday afternoon on the errand of patriotism which has twice carried him across the continent." The city treated the expedition well. Mayor William B. Thompson issued a permit to sell literature, there were no problems with lecturing, and Cora Mardon rejoined the expedition. On May 31 Ezra wrote Carrie that he was packed up to ship through the Great Lakes two hundred miles to Buffalo, New York. He added, "I can not see but my strength remains with me the same as five years ago."[12] Meeker was constantly amazed at his good health and mentioned it often in his correspondence.

Retreat from New York

The expedition arrived in Buffalo around June 2. Meeker planned to stay in the area for two weeks and then drive across the state to New England. He hoped to be in Rochester by June 20 and stay a week, and gave instructions to various shipping warehouses, Curt Teich & Company, and his many correspondents to send his mail, books, and postcards there. On June 10 the expedition was camped at Niagara Falls. As it turned out, that was as far east as they would go.

Two occurrences prevented the expedition from continuing on to New England. The first involved Hiram H. Edgerton, the mayor of Rochester, New York. What exactly transpired is not known, but it was serious enough to cause Meeker to abandon his plans for New England. On June 20 he wrote the mayor:

> Your words yesterday hurt. I am an old man and of that class known as the "winners of the farther west" of which but very, very few are left. In one's declining years it does hurt to have one of the later generation show their disrespect by harsh words and harsher actions particularly from such as are honored by their present fellows. I am past the age to hold a resentful feeling but one cannot entirely suppress a feeling of sorrow to be taken as like a faker and talked to as such.[13]

The next day he wrote Curt Teich & Company that he was abandoning plans to tour New England and that he was transferring to Chicago. He directed that all postcards and correspondence be sent there, and also wrote his daughter to send future mail to Chicago. Apparently Mrs. Mardon returned to her New York home as there is no mention of her accompanying her husband to Chicago.

The Ackerson Note

A second and more personal affront to Meeker was the receipt of a legal document from Carrie and Eben Osborne. It stated that Meeker had borrowed $1,000 from Mrs. Louise Ackerson to help finance his pioneer exhibit at the Alaska-Yukon-Pacific Exposition in 1909, that Carrie and Eben had co-signed the note, and they intended to stop the fifty dollars a month payment they made to Meeker for the purchase of the Puyallup mansion and divert it to Mrs. Ackerson until the note was paid in full. It also stated the $397.13 they were holding for Ezra would be turned over to Mrs. Ackerson.[14] It cannot be overstated the effect this had on Meeker. His dependence on that stipend and the reserve fund was absolute. Without it a risky trip to New England was not possible. While Meeker responded politely, he was seething inside. He replied at length in two letters in a tone that could only be described as stiff, disputing the amounts, stating Carrie and Eben should be holding $460 and Joe Templeton $451.[15] Meeker drafted a new document directing payment of these amounts to Mrs. Ackerson. He also mentioned that when the Osbornes bought the Puyallup mansion there was discussion of making some living arrangement for him in Seattle to the tune of twenty-five dollars a month and that it was his preference that it be added to the fifty dollars monthly he currently received. In other words, he requested a raise of 50 percent per month, a subject he had previously broached in a January 31 letter. He ended by saying, "I think such a contract should be acknowledged and witnessed; don't you, Eben? The only reason in the world why I want a written contract is to cover a contingency which might come that you should be called first."[16] After this outburst Ezra let the matter sit for a time.

A few months later his displeasure boiled over again. His expectation was, and had always been, that his children, relatives, and friends should support his cause and incidentally him. His habit was to borrow funds from Eben and Carrie as often as needed. He took advantage of

Joe Templeton's legal help without ever mentioning reimbursement. He had his teeth repaired by a relative, and his medical needs met by a grandson. Meeker also borrowed often from his historical society friends, William Bonny, Henry Hewitt, and Charles Hood. He gave little consideration to the sacrifices, personal and financial, that these people made on his behalf. He eventually paid back all borrowed money, but usually on his timetable. His grandson, daughter, and son-in-law now insisted he comply with their wishes.

In August Ezra blasted both Joe Templeton and Carrie for withholding from him the disputed $900 dollars. He had tried to arrange for the printing of a new edition of *Ventures and Adventures* in Chicago but was unable to produce the required deposit. Their point was that paying off the Ackerson debt for which they were co-signers was a higher priority, while Ezra argued that if he had been allowed to print the book, all or most of the copies would have been sold, the Ackerson note paid off, and a surplus put in his pocket.

On September 8 while in St. Paul, Meeker wrote a letter to his old friend and business partner John Hartman Jr. that he labeled "Strictly Confidential." He began the letter with some pleasantries but quickly got to the main point. "I must go to Washington in December or I will most likely lose the object I have been striving for. I will not go and leave those Seattle parties as endorsers on that [Ackerson] note."[17] Meeker asked if Hartman would loan him $200 payable March 1, 1912, so he could pay off the principal. Ezra thought he could pay off the interest accumulation before July and asked Hartman if he would be willing to loan that amount also. Hartman agreed.

Ezra's anger over this issue dominated his correspondence. The letters to Carrie in which Ezra vented continued into 1912. The break with Templeton grew more serious when Meeker learned that Joe had used some of the disputed money to reimburse himself for his time and expenses and could not pay Mrs. Ackerson immediately.[18] In 1910 the two men exchanged over sixty letters. In 1911 the number was down to six. In 1912 they exchanged but two letters.

From Sioux City in August 2011, Ezra wrote his daughter, "I enclose two letters from Hartman that are self-explanatory. I now enclose $700.00 to be paid at once on the note and now if there are not funds available in Seattle to pay the remainder then advise me how much more to send as I want the note canceled and sent to me."[19]

By Ezra's accounting he had already sent more than enough to pay the $1,000 loan. The year 1911 ended with yet more correspondence concerning the troublesome and still unpaid Ackerson note. To his daughter, Ezra complained that Joe was improperly holding funds that belonged to him.[20] Although he concluded that he had no more to say about the matter, he fired off a letter to Joe Templeton a month later in response to receiving a check for $250. "After the remainder in your hands is paid I want to cast the whole event behind us as far as possible in memory," Meeker wrote. "God forbid I should trump up any animosity against you, and even if I do not easily forget I will forgive and look forward to some day for a cordial greeting upon my return."[21]

On February 6, 1912, Meeker sent another letter to Joe Templeton venting his spleen, asking again why Joe held the money for a year rather than paying it toward the Ackerson note as he instructed, and that, according to his bookkeeping, Joe owed him an additional $176 beyond the $250 previously sent.[22] On February 12 an obviously miffed Joe Templeton wrote his grandfather disputing the claim. "I have done you many valuable services in connection with the A. Y. P. tragedy and feel that the $250.00 sent to Texas was very properly due me for my services, and more too."[23] Ezra received an additional $185.53 from Eben and Carrie, ending the Ackerson affair but not without hard feelings between all the parties involved.

Midwest Comfort

Returning to 1911, on June 30 Meeker was in Hammond, Indiana. He secured permits and arranged for a camping place for the outfit that was following a day behind. The expedition camped until July 3 in Hammond. The summer heat reached temperatures as high as 98 degrees. Meeker sweated profusely and at times felt dizzy, but his greatest misery was the flies that were attracted to the oxen. Fourth of July found Meeker in Gary, Indiana, making arrangements for the upcoming Iowa State Fair. From there it was on to Chicago and the usual permit battle. Meeker had trouble at first even getting an appointment to see Mayor Carter H. Harrison, but he prevailed and eventually got his permit. He remained in the area about a month where, among other venues, he lectured before the Chicago Historical Society. He also received an invitation from Mrs. Ellen Hartwig Kalley asking him to attend the first Centennial Celebration of the

Cumberland or National Road in Indianapolis, which included a banquet and entertainment. Mrs. Kalley and Ezra would correspond often about the future "Pioneer Way."

After four prosperous days in Rock Island, the expedition boarded the steamer *Helen Blair* and traveled down the Mississippi River to Keokuk, Iowa, where they were the guests of Mayor Joshua F. Elder. The following day (August 17) the expedition steamed up to Burlington, Iowa, a city they had visited five years earlier. Meeker and Mardon drove the wagon off the steamer and through town to a camp at the corner of 4th and Jefferson where large crowds greeted them. That evening Meeker gave a lecture from the wagon. He shortly thereafter departed from Burlington and moved west to Ottumwa, Eddyville, and the state fair at Des Moines. Interestingly Meeker made no mention this time of having briefly lived in Eddyville in 1851–52.

Arthur Robert Corey was the secretary of the Des Moines State Fair and the men exchanged several letters working out arrangements. Corey sent Meeker two banners advertising the fair to be attached to the wagon. He also informed Ezra that August 31 would be "Old Settlers Day" and all fairgoers who were Iowa residents when the state was admitted to the Union would be granted free admission. Corey went on to say, "[N]o doubt you will meet a number of your acquaintances who live in the southern part of the state and along the road you traveled years ago."[24] Meeker spent August 24 to 31 at the Iowa State Fair, and was finally making headway financially.

Meeker continued to hop from fair to fair. Next up was the Minnesota State Fair. Ezra had applied too late to gain entry, so he enlisted Edgar R. Harlan, curator of the Historical Department of Iowa, to write a letter on his behalf to the secretary of the Minnesota Historical Society seeking admittance. Harlan wrote a glowing letter of recommendation and the doors opened. The remainder of September was spent at fairs in Huron and Sioux City, South Dakota. On September 25 the expedition passed through Omaha, Nebraska, by train, on the way to Oklahoma and Texas to winter.

OKLAHOMA TO TEXAS

Meeker went south to Oklahoma and Texas in the fall of 1911 for two reasons. The first was to wait out the harsh winter months in a warmer climate. The second was the prospect of a winter campaign in which he could make some money as he had sent home all his reserve funds to pay off the Ackerson loan. Sales and lectures in the Midwest during the winter were likely to be sparse. He hoped the south, where the weather allowed people to be outdoors more of the time, would be a more lucrative venue.

Meeker's first stop was the Oklahoma State Fair from September 26 to October 11 where he met Joe Grimes of Kingfisher, Oklahoma. Grimes was the owner of two very large, five-year-old oxen. Meeker described them as "18 hands high weighing 6500 pounds (and no fat) probably the largest yoke of 5 year olds on the continent."[25] Ezra was entranced and took out an option from Grimes to try them out until December 1 and then to buy them. Thus the oxen, Dick and Harry, became part of the expedition in Oklahoma City and the new star attractions. Meeker wrote the firm of Nichols, Dean & Gregg in St. Paul describing the oxen and asking them to "Notify me as to the largest ox yoke you have in stock, the length, and the width between the bow holes and also length of bows.... Please quote price for yoke and for ox shoes."[26] The *Daily Oklahoman* on October 1 ran a long story about Meeker, his stay at the fair, and his work, complete with picture—but it was buried on page twenty-two. It was a harbinger of ill things to come.

Meeker in Texas in 1911 advertising the "Largest Yoke of Oxen in the World."
Author's collection

Meeker described the Oklahoma Fair as a congregation of shows rather than a true fair. He placed the "giant" oxen Dick and Harry in a tent along with Dave and Dandy, "the world's most famous oxen," and charged admission to view them. Ezra had the cattle weighed and discovered that their weights were much less than expected and considerably short of the 6,500 pounds advertised. Harry tipped the scales at 2,100 pounds and Dick at 2,206 pounds. Mardon and Ezra did not believe the numbers and had the scales balanced and even weighed themselves to check for accuracy. When Dave and Dandy's combined weights came in at 3,794 pounds as expected, Meeker concluded the scales were correct. He wrote Grimes saying, "This detracts largely from exhibition value as well as the salable value of the cattle and if you expect to do business with me to sell you would have to at least cut the price in the middle at least."[27] Nonetheless, on the back of the ten thousand postcards Meeker had made of the oxen he wrote, "These giant oxen…are believed to be the largest yoke of oxen in the world now."[28]

Mardon did not accompany Ezra when he left Oklahoma for Dallas, Texas. He disappeared for a time to join his wife in New York. Ezra never explained his absence and simply spoke of a new hired hand called Doc. Meeker had made his application too late to obtain a space at the Dallas Cotton Palace, so he set up camp at the corner of Roy and Commerce streets outside the fairgrounds. He invited the members of the Texas Press Association to visit his pioneer tent at any time during the fair. All members who visited received free copies of his booklet "The Lost Trail." He needed the publicity as sales of his literature had dropped off precipitously. Then Harry turned up lame. Ezra called in a veterinary surgeon who found, deeply imbedded in the ball of the foot, a tack that had to be cut out.

To Mardon he wrote, "Have not much encouraging to write about; things in the tent have resulted just a little better—$17.35 yesterday; tremendous crowds reported at the Fair Grounds today but I miss my guess if we go over $20.00 today; now comes worse that Shreveport will refuse me admittance. I enclose $10.00."[29]

On October 28 Ezra went to Fort Worth to get permits and promised he would return later with the ox team. The next day the *Sunday Star Telegram* ran a long article about Meeker and his work. He rented a room for a week and ended up staying until November 11. Ezra wrote the Fort Worth school director, Professor James W.

Cantwell, inviting the teachers and students of the city to visit his camp and view his outfit free of charge. He offered a free copy of the booklet "The Lost Trail" to each teacher who visited and proposed to lecture on the importance of perpetuating the history and experiences of pioneers. He concluded by stating. "We can admit 600 or 800 pupils a day if hours could be arranged so as not have them all in a bunch."[30]

The trip south was a financial disaster to this point. Meeker now had forty-eight fewer dollars in his pocket than when he was in Oklahoma City. The railroad had charged him $128 to ship from there to Dallas, a charge Meeker found outrageous but which he was unable to get reduced. In Fort Worth he was shut down for violating the Sunday law, was not allowed to camp, and forced to pay five dollars a day for a hotel room. His Oregon Trail lectures ended and he was primarily displaying Dick and Harry in a tent for income as he cast around for his next location. On November 12 Ezra shipped to Waco. While waiting for the train he wrote Joe Grimes telling him that while in Washington, DC, in 1907 on a visit to the Smithsonian Institute he decried to the superintendent the lack of a pioneer exhibit. The superintendent agreed that the criticism was just and the two men informally agreed that when Meeker was through with the oxen "the government can have them to mount and with the wagon make a permanent exhibit of pioneer ways and pioneer days."[31]

Ezra continued to send Mardon an occasional ten dollars toward the back pay due him, and told him not to hurry back. However in a November 19 letter, with the remainder of his back pay enclosed, Ezra told him he could use it to return to Texas and resume his old job, but suggested that if he had employment at home it might be safer to remain, as prospects were rather grim in Texas. Meeker also wrote Curt Teich & Company that he had no funds to pay his outstanding postcard bill but he hoped to remit soon. Money was so tight that Ezra even argued over the garbage bill from his stay in Waco. On December 5 the expedition was in Tyler, Texas, where Mardon rejoined the team.

December 13 found the expedition in Austin, and by Christmas Eve they had reached San Antonio and were camping within the walls of the Alamo. With permission of the state authorities, Meeker's tepee tent, cot, cook stove, and "grub box" were all set up inside that famous building. Meeker and entourage were the sole occupants. He experienced a small change in financial fortune here as Dick and Harry were beginning to draw large crowds. "[W]hen the people see them a short distance away they flock to the wagon and ask all sorts of questions and when too many clamor then I make a short speech."[32] He told Grimes he lost money when exhibiting the oxen in a tent, but when he hitched the four oxen to the wagon people ran to them, asked dozens of questions and bought his literature freely. Ezra celebrated his eighty-first birthday in the Alamo. At the end of the year he wrote Grimes about his concern for the health of the ox Dick who was not eating properly. The veterinary doctor examined and treated him twice. Still, Meeker was uneasy about him. Instinct told him the four oxen would be a huge draw when he returned to the trail, but he didn't have the money to purchase them and his concern about Dick's health made him unwilling to take the chance. He also mentioned Grimes' offer to purchase Dave and Dandy. Meeker suggested that he would consider it if Grimes would agree to preserve the hides when the animals died and ship them to the Smithsonian Museum.

Meeker and Mardon at the Alamo, December 1911. *Author's collection*

CHAPTER 8

The Second Expedition, 1912

Escape to Missouri

As the New Year dawned Meeker was told to vacate his cozy Alamo home by January 8. Concerned about Dick the ox not eating, Meeker cancelled his plans to go to Houston. With his finances low Ezra did not want to take the ox in any direction except closer to the animal's home in Kingfisher, Oklahoma, just northwest of Oklahoma City. The ox was showing signs of great distress, but was a puzzle to the veterinarian. Ezra shipped the outfit by train to Sherman, near the Oklahoma border, hoping to make one more stand in Texas. When the city authorities refused to issue a permit, Meeker kept on going. Giving up on his plans to purchase the big oxen Dick and Harry, he dropped them off at Kingfisher, and continued on to St. Louis. Ezra had family there. His cousin Jacob Meeker was pastor of Champion Hill Congregational Church and influential in local politics.

On January 17 Meeker checked into the Moser Hotel with thirteen dollars left in his pocket. The railroad charged $307 for shipping the outfit north. Since that was more money than he had, Ezra made a partial payment. The outfit was expected to arrive in a couple of days and he was eighty-three dollars short of funds to pay the freight bill. As was his habit, he borrowed the money, this time from his cousin. That afternoon Jacob Meeker accompanied Ezra to the mayor's office on a mission to obtain the ever-elusive permit to sell his literature on the city streets. He was warmly greeted, but the wheels of bureaucracy ground slowly and the permit took a while in coming. Ezra next arranged with local theaters to give lectures for fifteen dollars a day. Bad luck seemed to be his companion. One afternoon Meeker fell asleep while reading in a chair in the hotel lobby. He was startled awake by the sound of his glasses shattering as they hit the floor. Ezra paid five dollars for them in Seattle and was quite perturbed over losing them. Fortunately he found a much less expensive pair nearby. A week later his pocketbook was down to $2.80, but he and Mardon were out on the streets selling. Within days they were a week ahead on room (seventy-five cents a night) and board (fifty cents a day), and could breathe a cautious sigh of relief.

On January 29 Meeker gave his first St. Louis lecture to an audience of six hundred and had lined up lectures for the next several days. When they weren't selling, Mardon was out making arrangements with theaters for future lectures. Eventually the outlook improved. On February 4 the *St. Louis Dispatch* ran a large story about the expedition with many pictures. The check for $185.53 from Eben and Carrie helped to ease the strain.

Despite his struggles Meeker never lost sight of his mission. He wrote Congressman Humphrey almost immediately upon arriving in St. Louis asking if there was any chance of getting a hearing for the newly introduced bill in the upcoming session of Congress, or if it would be better to wait until after the presidential election. Humphrey told Meeker, "I think if you come to Washington City that a hearing can be arranged for you without much delay. What will be the result I cannot undertake to prophesy."[1] The next day he advised him not to come. Ezra sent the congressman the seventy-foot Nebraska portion of his giant Oregon Trail map and asked, yet again, if he should come to Washington. Humphrey replied that while Meeker's presence would be helpful, he thought he could get the bill out of committee without his aid.

By the middle of the month Ezra had arranged with McAdoo & Company of St. Louis to print ten thousand copies of a new but smaller paperback version of the *Ox Team* under the title of *Personal Experiences on the Oregon Trail*. The price was five and one half cents a copy. He wrote a long letter to Curt Teich & Company explaining his financial difficulties and promising a remittance soon. Elizabeth Gentry of the DAR came to his rescue. She sent Meeker an invitation to participate in a "Land Show" in Kansas City with all his expenses paid, including the cost of rail transport ($110) from St. Louis.

On February 24 Ezra took the train to Kansas City. Mardon and the team followed the next day. Meeker was still fine-tuning details of his new book. He made some changes in the text and reduced its size, doing the business by letter from Kansas City. Elizabeth Gentry's name and address were added to the book jacket, listing her as a contact person from whom to order copies, suggesting that the DAR may have helped with the financing of the book. By March 7 he was taking delivery of the books. Meeker also arranged for another printing of twenty thousand copies of his booklet "The Lost Trail." Then he apologized for his late payment to McAdoo & Company saying, "I have been held up first by the snow and now by the flood, constricting my sales while yet adding to my expenses, yet I think I can soon send you further remittances from my sales here."[2] The usually optimistic Meeker told Carrie, "I am well but must confess much depressed in spirit."[3] Finances had much to do with his depression. He wrote Curt Teich, "Well, my good Friend, have patience and in a little while you will hear glad tidings."[4] He was delayed by a spring snowstorm, but on March 12 the expedition shipped by rail to Iowa.

Iowa and Nebraska, One More Time

Upon his arrival in Council Bluffs Meeker checked into the Ogden Hotel. Mardon stayed with the wagon and team. Financial woes continued to be a constant, nagging weight. Curt Teich & Company wrote, "We are very sorry to hear that your trip to the South was not successful, and you have been losing, but as we need the money, and can use all we can get at the present time in order to pay our bills for the new factory, we surely would appreciate it if you could send us a check on account."[5] To W. S. McAdoo & Company Meeker wrote, "The snow has completely blocked my business and I have no mail from home for a fortnight, hence cannot send you remittance today but will remember you soon."[6]

By the end of the month Ezra had moved across the Missouri River to Omaha and was staying at the Boquet Hotel. On March 25 and 26 he delivered an illustrated lecture, "Life on the Great Plains 1852," in the Besse Theater in South Omaha for the sum of thirty-five dollars. The last three days of March found him lecturing afternoons and evenings at the Cameraphone Theater for fifty dollars. As in past practice, Ezra agreed to advertise for theaters by placing banners on the wagon. Just as he was beginning to climb out of his hole, he lost Mardon.

End of the Partnership

Shortly after arriving in the Council Bluffs/Omaha area, William Mardon left the expedition for the final time. He joined his wife in Johnstown, New York, and soon thereafter the couple moved to Stamford, Connecticut. Apparently the parting wasn't totally amicable. Ezra told Carrie, "Mardon has left me to go to New York to get a divorce—says his wife is drinking; I have finally succeeded in nearly paying him off though it wasn't the square thing to draw high pay for winter when but little could be accomplished and leave when spring opened."[7] The bad feelings ran both ways. Mardon asked Ezra for all his back pay. Meeker didn't have the funds and offered to pay in installments, at which point Mardon lost his temper and threatened to put a lien on the oxen and wagon if he didn't get his money. Despite the acrimony Meeker sent Mardon off with a brief and somewhat begrudging letter of recommendation.

> The fact that you have been in my employment for over five years and left it of your own notion for family reasons and that I have offered you the opportunity to return to your old place in the event you decided in the near future to do so, is, in my opinion, as thorough a recommendation as I can make and you are at liberty to show this letter to whomsoever you may choose.[8]

Shortly thereafter, Cora Mardon's mother, Ethel Miner of Johnstown, New York, unexpectedly weighed in on the matter.

> I felt very bad about Mr. Mardon leaving you when he did.... It was not his fault, however. He would not have left you but for his wife. She was constantly urging him to come home and also made demands on his purse to such an extent that he was at his wits end to gratify her wishes.... My daughter is very unreasonable at times. I know.[9]

Ezra replied telling Mrs. Miner he held no ill feelings toward her son-in-law. He ended saying, "I sympathize deeply with him, and yourself for the troubles that confronts you but we all have our perplexities in life which we must need bear in patience, submitting to the will of the Father who in his good time will call us to the better and happier life."[10]

The two men exchanged a few letters over the following weeks. Mardon wrote asking how things were going. Meeker wrote a little each time he sent Mardon a portion of his pay, telling him of the new man and wife he had hired, how sales were going, and mentioning the "motion picture scheme" of crossing the Loup Fork River in his wagon box. A puzzling line appeared in one of Ezra's letters. "Nothing is the matter with Teddy [Roosevelt], 'he's all right,' I heard him in the big auditorium with no standing room left. He is indeed a great man."[11] This

Meeker caulking his graffiti covered wagon prior to filming at the Loup Fork River on May 23, 1912. Note the graffiti on the wagon. *Ox-Team Days on the Oregon Trail*

encounter with President Roosevelt would have been the third time the two men's paths crossed, but when and where it happened is not known.

Mardon quickly secured a job as a conductor, later motorman, on the Connecticut trolley system in Stamford.[12] The transition from ox team drover to trolley driver must have been a natural one, as he turned this new job into a lasting career. He and Cora divorced in 1921. Both eventually remarried. William died on February 6, 1948, and is buried in Stamford's Fairfield Cemetery.[13]

Heading West

Snow, high water on the roads, and washed out bridges pinned the expedition to Omaha for a time and prevented driving to Grand Island. However, Meeker's lectures provided some compensation. A new helper named Judson Barrett was hired at two dollars a day. His wife, whom Meeker described as always cheerful and competent, and ten-month-old baby girl, Helen, who was "good-natured, plump and handsome," came along also. Barrett was initially described as "not quite as good a man" as Mardon, but by the end of April Meeker's opinion had changed saying, "I have help that suits me decidedly better."[14] Mrs. Barrett was not paid wages and Meeker made no promise as to how far she could go.[15] Judson Barrett was a photographer with his own camera and while in Omaha Meeker supplied him with photographic material including plates for making slides and a small folding dark room where Barrett could develop the photographs. Meeker hoped snapshots of towns and people along the way could be developed instantly and shown that night, giving each lecture some local spice. He also envisioned people paying to have their pictures taken with the ox team and wagon.

By the end of April Meeker was ready to leave Omaha. The snow had melted, the bridges were repaired, and the roads were now passable. On May 1 Ezra started the team west under the supervision of the Barretts and went ahead by rail to make arrangements for future lectures. He turned his stock of literature over to the Barretts to sell, writing his daughter that he was certain they were honest people. Nevertheless, Ezra counted his stock and receipts often to make sure the books balanced. The next planned stops were Waterloo, Fremont, and Columbus. Once on the road Meeker announced, "I slept in the tent last night, the first time for about six months and rested fine on my cot…. I enclose postal card of baby Helen; her dress is red which with her sharp black eyes makes a striking picture."[16]

Motion Pictures

In a letter to his daughter shortly before leaving Omaha, Meeker made mention of showing a motion picture taken in the area to an audience of nine hundred.[17] The motion picture business was in its infancy during the Old Oregon Trail Monument Expedition years. It did not take Meeker long to see the potential of the medium for advancing his cause. While in Cincinnati in February 1911 he wrote Orville and Wilbur Wright in nearby Dayton, told them of his work, sent them some newspaper articles, and said,

> I can arrange to have a moving picture of my team taken in Dayton and which would be very helpful to me could I but have with it some other attraction to show with my lectures. I write to ask if you would arrange so we could get a flying machine in flight, just rising and again returning or soaring—the sure enough thing. I would like to meet and confer with you and show you further about my work.[18]

There is no evidence that the meeting or the filming took place, but if such a film exists it would be priceless.

In Dayton, the Cash Register Company financed the making of a very short motion picture clip of the ox team in front of their factory. Ezra used the film at his lectures and begged for more. "I have received the film and it is fine, but, like a scrimped dinner there is not enough of it and leaves a longing for more." By the time Meeker reached Troy, Ohio, on February 3 the short motion picture of the ox team had become a lecture fixture. In an April 12, 1911, letter he suggested that the film include an introduction to read something like "Ezra Meeker's Oregon Trail Monument Expedition at the Cash Register Works, Dayton, Ohio or Ezra Meeker addressing employees of the Cash Register Company, Dayton, Ohio." He also suggested making the movie longer by filming some scenes when he arrived in Toledo on April 17, touting the advertising it would offer to the two thousand people he lectured to almost daily. Apparently, he was able to convince the Cash Register Company that a longer film would get him into the more important theaters.[19]

He also wrote the governor of Washington asking for some slides of the state to incorporate into his lectures and attempted to obtain a film clip of the team taken in 1909 at a fair in Spokane. Meeker used these motion pictures in the many lectures that he delivered over the next couple of years.

In March 1912, while traveling between Kansas City, Missouri, and Council Bluffs, Iowa, Meeker lost the film. He wrote to just about everyone he could think of trying to locate it and in doing so gave a hint as to its length. To the Rock Island Railroad he wrote, "A moving picture film in a round tin case is missing. My driver thinks I left it at your hotel or in the room; it's about a foot in diameter and about three inches thick. If you find it send by express without delay to Boquet Hotel, Omaha Neb."[20] Eventually the missing film canister was found and returned.[21]

An even longer film was soon made. In April 1912 while in Omaha, Nebraska, Ezra met a motion picture artist named Oscar A. Albrecht of Grand Rapids, Michigan, who proposed making a feature length film of the Oregon Trail. It was to be "4000 feet of moving pictures of the ox team on the way in six states including Denver, Salt Lake, and the grand scenery down the Columbia river."[22] Meeker was to deposit $500 before the work started. Albrecht estimated that the project would take fully three months to complete.

Of course Meeker did not have the money. Still, the idea was too tempting to pass up. He contacted Joe Grimes and proposed a business deal. Grimes was running a Wild West show in Oklahoma and Ezra offered him a copy of the film when it was finished, which he said cost $1,000, if Grimes would advance $500 immediately. Meeker waxed eloquent about how the film would be the perfect addition to Grimes' Oklahoma performances. Ezra promised Grimes the exclusive rights to show the film in Oklahoma and even offered Dave and Dandy as security if he failed to deliver the film.[23]

He also wrote to Charles Burton Irwin of Cheyenne, Wyoming, who operated a Wild West show in that state.[24] Irwin offered to fund the project if Meeker joined his show and perhaps asked for full rights to the film. Meeker countered and asked for the right to show the film when he lectured saying,

> One thing is sure the opportunity will not offer again for myself as I am now past 81 years and I do not know of another pioneer willing or physically able to undertake the work; besides literally millions of people have seen my outfit in the east & middle west and over a thousand newspapers have given extended notices of my work; I send you one where 298,000 copies were printed.

He further told Irwin, "[N]ow have 400 feet of film & slides showing the approach to Kanesville (now Council Bluffs) as also as it appeared to me 60 years ago, and street scenes of Council Bluffs."[25]

He wrote Carrie, "I am now negotiating with an artist here to take 4000 feet on the Trail and have hopes we can accomplish it and if I do, then good by to the ox team as I can get a hearing anywhere without it. If I succeed in this, the work will extend down through the Columbia river gorge to Vancouver."[26] Goodbye Mardon, goodbye ox team? Never daunted by change (or excess sentimentality), Meeker tilted toward and embraced the future.

Meeker also lobbied the railroad companies to help with the expense. He contacted both the Union Pacific and Oregon Short Line to gauge their interest in the project, but made no headway.[27] By the end of April he reported to Carrie that he was showing some new footage.[28]

Using his own funds, Meeker had doubled the amount of film footage and told his daughter in early May of a plan to take more.

> My new moving pictures are fine indeed and I now have over 800 feet of film; now don't go and get uneasy when I tell you I am planning for a moving picture scene of crossing the Loop [sic] river in my wagon box, for I will arrange for one or more boats to "stand by" and tow me in the event the current is too strong for me. The picture, if I succeed as planned will be a striking illustration of early pioneer life on the Plains and will enable me to get many engagements to lecture.[29]

On May 19 Meeker was at Columbus, Nebraska, preparing for the filming. He described the proposed river crossing as being one hundred rods in length and that he intended to tie up in the middle of the river while the cameraman changed to the opposite bank. Ezra assured Oscar Albrecht that he would be ready when he arrived and suggested that he might also be able to ford the nearby Platte River.[30] He arranged with William W. Scott to take snapshots of the scene. All went well as he wrote Carrie, "Yesterday accomplished the crossing of the Loup river in my wagon box and will soon have fill of the moving pictures of the scene as also slides of 14 snap shots; it is quite an event."[31] That Friday he was back in Grand Island showing the newly taken film.

On July 2 Meeker was in Fort Morgan, Colorado, where he wrote Oscar Albrecht he intended to remain until July 6 before moving on to Denver.[32] He requested editing one short scene of the Loup Fork crossing and wanted additional footage taken to double the size of the film to 1600 feet. He also sounded out Albrecht about filming in Denver.

Albrecht responded that he could film another 800 feet for $185 if the filming could be completed within three days. Anything beyond that would incur an extra ten dollars a day surcharge. He told Meeker as soon as he wired a hundred dollars he would order the film from New York and come to Denver as soon as it arrived in Omaha.[33] It was more than Ezra could afford. He responded from Denver, "Your letter of 16th reached me this noon. I would not care to incur so large an outlay as named in your letter for 400 or 800 feet of films respectively; in fact just now couldn't."[34]

Where are these motion pictures today? The section showing a scene of Ezra driving the wagon through a city street and scenes of the crossing of the Loup River survived. A copy was found in Meeker's wagon, which at the time was on display at the Washington State Historical Society Museum in Tacoma, Washington. The Puyallup Historical Society at the Meeker Mansion also has a copy. A third copy with footage of Meeker reading to children, along with the Loup Fork crossing, is in the Howard Driggs Collection at Southern Utah University in Cedar City, Utah. The fate of the remainder of the film is currently unknown. A decade later Meeker renewed the effort to make a feature length motion picture about the Oregon Trail.

July 3 found Meeker dabbling again in new technology. He made a phonographic recording of a speech he had been asked to deliver a week later at the annual gathering of the Pierce County Pioneer Association in Puyallup. The occasion was the dedication of a tablet honoring the Meeker's first home in Puyallup. Ezra mailed the recording from Fort Morgan, Colorado, to Puyallup and it arrived in time to be played for the gathering of some 1500 pioneers and Meeker relatives. Unfortunately, the sound system was inadequate for anyone standing a distance away to hear, and his niece had to read the speech aloud to the audience. The fate of that recording is unknown.

HOME STRETCH

By the middle of July the expedition reached Denver just in time for yet another disaster. A flash flood devastated the town and most of Meeker's stock of books and postcards (a thousand books and 30,000 postcards) got soaked in the downpour. Meeker worked patiently to dry and clean them but found soiled literature would not sell. He wrote, "Even if I had good stock the public mind was so upset by the appalling disaster of the cloud burst nothing else was talked or thought of but to relieve the sufferers by the flood."[35] He offered city officials free of charge one of three illustrated lectures, "The Lost Trail," "The true story of the Oregon Missionaries," or "Sixty years on the Frontier," all fully illustrated with stereopticon views and special moving pictures as a way to help a suffering populace take their minds off of the recent floods. He scheduled a lecture at the Ivy Theater, but the plan was thwarted by a Mr. Buckwalter, agent of the General Film Company, who contacted both the theater management and Ezra claiming his company had exclusive rights to the showing of movies in the city's theaters. Ezra tried to meet with him to iron out the problems, but failed. Meeker sent Buckwalter a letter on July 21 explaining what he was doing, and in conclusion cited the dozens of officials who had opened their cities to his project. He even cited the support of President Roosevelt, which was a bit of a stretch. He asked Buckwalter to meet with him personally. Instead Buckwalter responded with a letter that enraged Meeker, and led to this over-the-top reply:

> I have your letter of 23rd in which you say "we have no jurisdiction whatever over the matter of license and unlicensed pictures." Then why did you object to my pictures being put on at the Ivy theater as you did to me over the phone and afterwards to the theater management? If you have written the truth in this letter before me, then you have laid yourself liable to me for damages as well as for criminal prosecution for assuming to act for others without authority and to the damage of third party. I have known men to go behind prison bars for just such an act. You ought in very shame hide your face from civilization.[36]

The Denver experience sent Meeker packing. The team left town the night of July 23, heading to Golden, Boulder, and Longmont before Meeker gave up entirely on Colorado. Stress was taking its toll. Newspapers in Castle Rock, Washington, and in Salem and Ashland, Oregon, reported an incident where Meeker physically assaulted a man in Longmont on August 5. According to the story, the ox Dandy had come up lame, and George Bashor, claiming authority as a humane officer, told Meeker Dandy wasn't fit to drive. Meeker replied that he considered it none of Bashor's business. Bashor then pulled his buggy in front of Meeker's team and informed him that he was under arrest. Meeker responded by lashing out at Bashor with the long blacksnake whip he was holding. The newspapers described the scene: "When Meeker got the whip loose he went after Bashor again and did not stop until he had again wrapped the whip around his neck and broken it in pulling it loose. After investigation it appeared that Bashor had acted without authority and the matter was dropped."[37] The story seems out of character, even for Meeker. His more familiar recourses were wars of words or legal procedures. Also he was 82 years old at the time and had never been noted for dexterity with a whip. One suspects events were dramatized for effect. Nonetheless, it was clear Meeker needed to wind down his expedition and head for home.

Meeker shipped the next day to Cheyenne, Wyoming, to participate in the city's Frontier Days celebration. While he was asleep on the train, the dog Jim somehow escaped. Ezra sent a man back along the tracks thirty-five miles in a futile search for him. A second search the next day led to the person who had found Jim and now didn't want to part with him. After quite a "rumpus" (in Ezra's words), Jim was returned. Ezra wrote, "I did not realize I thought so much of Jim until for awhile I thought he was lost for good and all."[38]

Meeker planned to ship the outfit from Cheyenne to Seattle, anticipating arrival home at the end of August. He wrote Carrie that he had chartered a boxcar for $200 and assured her that he would be comfortable in the boxcar with the oxen and his dog Jim. He then asked Eben for a loan to cover the $200 freight charge, since he only had sixty dollars at hand. He pledged to mortgage the oxen, wagon, and stock of books and postcards and to go to work selling as soon as he arrived. He hoped to have the loan repaid within ten days. He then wrote Charles Hood of Puyallup, reciting his woes, and asking if he would arrange a lecture soon after his arrival.

> I have fought a strenuous battle but have met defeat; the particulars I will not undertake to write, but I am going home without any accumulation or money to pay the freight; suffice it to say I lost nearly my whole stock of books I had with me in the Denver cloud burst and now have lost the use of my fine ox Dandy by founder and may lose him entirely and anyway can't do any business here without a full team. I have replenished my stock of books & cards from those I had stored in Chicago & St. Louis and will ship my outfit to Puyallup about the 28th.[39]

Meeker fretted that he was going to lose Dandy and decided to unload the ox at Puyallup as it would be easier to find a more suitable pasture there than in Seattle. He did not know if Puyallup had a movie theater but suggested if so he would show his film if he could get a projector. His long-suffering son-in-law sent the $200. Meeker thanked him, and said after Puyallup he would have Dave the ox pull the wagon alone to Seattle where he would continue to lecture and speedily repay the loan. On August 21 Meeker received a statement from W. S. McAdoo & Company showing he still owed a balance of $208.50 on his account.[40]

One final misfortune befell the bedraggled expedition. On the way home, on the night of August 19–20, Jim vanished again from the train somewhere east of Kemmerer, Wyoming. A search and an offer of a reward failed to turn him up.[41] Meeker tried hard to find Jim and his correspondence contains several letters to various people requesting aid in the search, but he was unsuccessful and his faithful companion for the last six years was truly lost.

No Help From Congress

On August 7, 1912, Meeker received tidings from Senator Wesley L. Jones that Senate Bill 5009 authorizing the appointment of a commissioner to locate the general route of the Oregon Trail and to supervise the erection of markers and monuments had finally passed in committee and had been sent to the Senate floor.[42] On August 15 the bill passed the full Senate. But House Bill 5966 was still pending and Meeker's lobbying continued. He urged Congressman Humphrey to push the effort to get his bill passed.

> It is highly important this survey should be made and made now. The people are moving 'Heaven and Earth' to push work on the southern route [Lincoln Highway], while nothing is being done on this route in question. Even if we could get only an appropriation to locate the Trail by actual survey and secure data as to favorability of the route and approximate cost of a highway it would be a long step in furtherance of the greater work.[43]

Ezra ended the year in residence at the Donnelly Hotel in Tacoma. From here on December 28 he wrote John Hartman Jr. that he had received notification from Congressman Humphrey that the bill had stalled in the House of Representatives. He went on to say that his next plan was to get the various state legislatures to send a memorial to Congress urging the passage of the Oregon Trail Bill. He told Hartman that he was on his way to Olympia to "try and pass it through."[44] Despite all the setbacks Ezra Meeker was not about to give up. The work, and Meeker, would go forward for another sixteen years.

CHAPTER 9

Interlude

Meeker spent the two months following his return home at various western Washington county fairs, selling his books and postcards, and paying down his debts. In late September 1912 he got involved in local Puyallup politics. When he originally platted the city in the 1870s he named three of the primary streets Pioneer, Meeker, and Stewart. In 1912 the Puyallup City Council changed the names to First, Second, and Third Avenues. Meeker protested and wrote a letter to the council requesting a hearing on the matter, and sent a copy to the local newspaper. "I care but little about the change of the name of Meeker street, but I do care a great deal as to Pioneer avenue."[1] The Puyallup Ladies Club took up the issue and collected signatures on a petition to the council asking them to restore the original names. Meeker sent them ten dollars to help with costs and promised more if needed.[2] The petition was enough for the council to put the question to the city voters. The tally to restore was 1042 for and 149 against.[3] Today Stewart, Meeker, and Pioneer streets continue to remind the Puyallup residents of their heritage, just as Ezra intended.

At the end of the year Meeker published another book—*Uncle Ezra's Short Stories for Children*. The slim volume, published in Tacoma, was a collection of short tales from Washington Territory pioneer days, aimed, as the title said, at children. The chapters described incidents such as a community Christmas celebration and his daughter's encounter with a mountain lion. Meeker never commented on his inspiration for this change of pace, but it joined his collection of literature for sale at his appearances and speaking venues. The *Seattle Post-Intelligencer* gave it an excellent review.[4]

The Oregon Trail was not forgotten, but Meeker's "feet on the ground" campaign was put on hold for the moment for lack of funds, if not for lack of ideas and energy. A slight change of emphasis at this

The cover of Meeker's 1912 book, *Uncle Ezra's Pioneer Stories for Children*. D. Larsen *photograph*

point widened the scope of his challenge. At the end of the year he wrote Senator Jones urging a change be made to the recently passed Senate Bill 5009. He told Jones, "I have from the start advocated adopting the Oregon Trail as the route for a national highway, but by the advice of Mr. Roosevelt omitted reference to it in the bill as he thought the time was not ripe for such a movement then." Meeker requested the bill now be amended to include funds for a survey of the route for a national highway. This needed to be done even if it became necessary to "drop the question of planting markers except something temporary."[5] Jones acquiesced, as did other members of the Washington delegation over the years.

The Fate of Dave and Dandy

In August 1912 the ox Dave arrived home healthy, but Dandy did not. The oxen were initially placed in a stable in Puyallup while Ezra mulled his options. He originally thought to donate them to the Smithsonian Museum, but only after their deaths. As they were still very much alive, he needed to look elsewhere. For a time he considered donating them to the Washington State Historical Society. He had roots there as a charter member and past president, and had a good working relationship with the current secretary, William Bonney. But by October 1912 he came to the conclusion that donating them to the City of Tacoma, and in particular the Metropolitan Park Board, was the best option. Point Defiance Park was large. It had a buffalo paddock and other pastures that would be suitable for keeping and exhibiting the oxen until their deaths. On October 27 Meeker formally brought the plan to Henry W. Meyers of the Park Board. He envisioned an exhibit to accompany the oxen and wagon that consisted of his camp kit, mess box, his homespun suit, two hats, his tent, and over one hundred photographs, all enclosed in a log cabin structure that would also include a large map of the Oregon Trail etched in glass.[6] He also hoped to include his dog Jim in the exhibit—if he could be found.

On November 12, 1912, the outfit was donated to the Park District via a legal document.[7] The understanding was that the oxen would be mounted when they died and exhibited in a thirty-two by fourteen-foot glass case as a permanent historic exhibit. On February 19, 1913, Dave and Dandy were moved from their Puyallup stable to Point Defiance Park and placed in a pasture next to the buffalo paddock where they were to graze contentedly until their end.[8]

The park board informed Ezra in June that the actual construction of the Point Defiance exhibit building could not begin until after passage of the fall levy. In the meantime they reported, "The two oxen who are the sole occupants of one of our large pastures are looking very fine and are attracting much attention."[9] But Meeker was apparently not quite ready to give them up. In August he asked the board if he could take the oxen on an excursion to Walla Walla for a local event, and by October he envisioned shipping the entire outfit to San Francisco for the upcoming Panama-Pacific Exhibition.[10]

The oxen had become northwest celebrities in their own right, and Meeker and the park board were concerned about a public outcry if they were euthanized rather than allowed to meet a natural end. A notice to be published in the *Tacoma Ledger* was prepared in advance for Dandy's death, and another notice for Dave.[11] This was premature, for at the beginning of 1914 the oxen were alive and somewhat healthy. The Park Department agreed to loan Meeker the outfit but insisted that if they died in route to San Francisco the hides be sent back to Tacoma for taxidermy. Further, they said any funds received for the meat must also be accounted for.[12]

A sketch of the proposed Point Defiance glass house. *Author's collection*

Dandy made it only as far as Portland. On June 17, 1914, he was slaughtered.[13] The hide and bones were shipped to Tacoma, and the meat was accounted for to the penny ($75.20, or $1,813 in today's currency).[14] Claude W. Du Bois was the taxidermist. Dandy's hide arrived in Tacoma infested with maggots, which resulted in a number of back and forth letters on how to properly prepare Dave's hide when the time came. Some thought Dandy could not be mounted, but Du Bois was able to complete the job successfully. On November 5, 1914, Dave met his end in San Francisco. Meeker remitted $74.50 ($1,796 in today's currency) for the meat and on November 11 shipped the hide to Tacoma via Wells Fargo & Co.[15] Since no mention was ever made of Dave suffering from ill health or conditions of any kind, it can be assumed he was humanely dispatched by Meeker to make a matched pair for exhibit at the exposition. Deprived of their retirement pastures, Meeker's loyal oxen were put to work in his service even in death.

At Death's Door

Meeker often boasted that he had never been ill in his entire life. When he said this he was usually commenting on his remarkable health and stamina as an elderly man, but he was not being entirely truthful. Throughout his life Meeker suffered from what he called indigestion. His family letters are full of complaints of stomach and intestinal disorders. In all probability what ailed him was a hernia dating from a log-rolling incident in Indiana in 1851 that plagued him thereafter. His solution, which offered relief, was to fast for a period of time. For over sixty years he wore a truss, a surgical appliance designed to support a hernia, typically a padded belt. On November 18, 1913, the tough muscles in his abdominal wall clamped down on a section of his intestine poking through it (the hernia) and squeezed off the blood flow. This condition is called a strangulated hernia and immediate surgery is required to relieve the pressure on the bowel, push it back inside the abdominal cavity where it belongs, and repair the tear. Speed is a necessity, as failure to act quickly can lead to death. For a man approaching his eighty-third birthday, this was serious business.

Meeker was in Centralia dealing with a family legal matter involving his youngest daughter and her aunt when he was stricken. Ezra was placed on a train and rushed to Providence Hospital in Seattle.[16]

A number of friends, along with his daughter, Carrie Osborne, her husband Eben, and his grandson Dr. Charles Templeton, met Ezra at the station and took him to the hospital where Dr. Templeton and two others performed the surgery. Meeker remained in the hospital over a month, mostly confined to his bed. However, bed confinement did not mean idleness. There are a number of letters in his papers dated during his time in the hospital dealing with such things as holly sales, legal matters, arranging for the Bon Marché department store to sell his *Children's Stories* book, and discussing the possibility of his attending the scheduled fortieth anniversary celebration of the first Northern Pacific Railroad trip from Tacoma to Kalama, an inaugural ride in which Meeker had participated.[17] Ezra regretted he was unable to speak as scheduled at the Washington Good Roads Association convention in north Yakima. A letter to his granddaughter Olive Osborne tells how the recovery went.

> I have submitted myself to a rigid vegetable and fruit diet, eschewed meat and kept my system in good working order and now the wound of the operating table is nearly healed as Charley says "like as if you had been a young man." All the same Charley saved my life; he thinks I will be better than ever when once thoroughly healed. The memory of that hateful truss of sixty years is enough to give me the horrors and if I get rid of that I will feel amply rejoiced and Charley thinks I won't have to wear it any more.[18]

Word of Meeker's brush with death appeared in the New York newspapers and prompted a long letter of praise and concern from Mardon's mother-in-law along with a few succinct comments, "I am glad and also surprised at your recovery from such a serious illness at your age. You certainly have a wonderful vitality and it may be that God will spare you years yet. It seems your work is not yet done."[19]

The surgery took its toll, however. Meeker wrote his nephew, "I am reduced in weight to 120 pounds (I was 165 when you knew me) yet feel strong and active, take long walks and otherwise exercise and have good appetite but with less digestive powers."[20] He never did regain the lost weight and shortly after the surgery suffered from a severe cough that for a time looked serious. But as the next fifteen years of his life demonstrated, his energy and drive to be productive remained.

Meeker, the Camp Fire Girls, and the Oregon Trail Memorial Association

In preparing for the 1915 Panama-Pacific Exposition, Meeker remembered back to 1870 when he made his first trip to New York City. He had 2,500 copies of a booklet touting Washington Territory that he hoped to sell in that city. He also brought with him a display of fifty-two varieties of pressed winter-blooming plants. Upon arriving in New York, Meeker sought and received an interview with Horace Greeley whose *New York Tribune* newspaper he had read religiously for eighteen years. Greeley examined the flower collection and suggested that it be shown at the New York Farmers' Club. A story about these wondrous winter-blooming plants was published in Greeley's newspaper, picked up by others, and eventually reached two million readers.[21]

Meeker decided to repeat the flower exhibit for the Panama-Pacific Exposition, but in a unique fashion and on a much grander scale. Working with Seattle Boy Scouts head Edward S. Ingraham (also the first superintendent of the Seattle Public Schools and a noted mountaineer who climbed Mount Rainier thirteen times), Meeker developed a plan to collect and press hundreds of varieties of summer flowers gathered from Paradise Valley, elevation 5,400 feet, in Mount Rainier National Park, in late July or early August, at the height of the bloom. The flowers would then be displayed "under glass" in the Washington Building at the San Francisco Exposition. The flower gatherers would be Seattle Camp Fire Girls.[22]

Members of Seattle's five Camp Fire groups would journey by train, automobile, and then on foot to Paradise Valley where they would camp in tents and, under the guidance of a professional botanist, collect and press the 360 varieties of flowers and ferns that grew on the flanks of Mount Rainier. The outing was successful. Meeker wrote the exposition president an account of the expedition.

> By a little publicity through the columns of the local papers, I set the ball in motion that resulted in an expedition to "Paradise Valley," up 6000 feet on the slopes of that grand old mountain "Rainier;" of 75 Camp Fire girls with the "time of their lives" where they made selection from over 500 varieties of flowers growing, to beautify in pressed form, the walls of the ladies rest room at the Washington State Building at the Exposition.[23]

Meeker included the story of the Camp Fire Girl's trip to Mount Rainier, complete with slides, in his lectures in the Washington State Building at the Exposition. It is not known if the exhibit went into the women's restroom where it would be unavailable to male viewers. In later correspondence Meeker stated that after his lecture and slide show about the outing many in the audience went upstairs to view the flowers.[24] There is no mention anywhere as to how the National Park authorities viewed the outing. Both camping at Paradise and picking any vegetation in the park would be strictly illegal today.

Meeker could not have imagined that his support for an outing of a group of Seattle Camp Fire Girls would eventually lead to the founding of the Oregon Trail Memorial Association (OTMA). Eight years after the outing, in 1922, Meeker's publisher Caspar W. Hodgson introduced him to George Dupont Pratt, an early leader of the Camp Fire Girls and a founder of the Boy Scouts of America. The two men found a common thread in their Camp Fire experience and Pratt listened as Meeker told him of his Oregon Trail efforts. Pratt was enormously wealthy, set Meeker up in New York City, began to financially support the Oregon Trail work, connected him with other men of means, and four years later in 1926, OTMA was born. Without the common thread of the Camp Fire organization it is likely that Meeker and Pratt would not have gone beyond polite conversation on the occasion of their introduction.

THE FINAL DRIVE

When Meeker secured the loan of Dave and Dandy from the Tacoma Metropolitan Park Department and started south in May 1914 for the Panama-Pacific Exposition, he actually had in mind a much grander plan to begin an eight-month long publicity campaign, taking the outfit by train to stops in Salt Lake City, Denver, Chicago, New England, and Washington, DC. The goal was to arrive in the nation's capital on November 29, 1914, and carry President Wilson from the White House to the capitol in the wagon (presumptive on Meeker's part) where Wilson would deliver his annual address. This, coincidentally, would be the seventh anniversary of Meeker, wagon, and ox team meeting President Theodore Roosevelt in Washington.[25] All this, of course, was fantasy, dependent upon his health, the health of his oxen, and the exposition, the railroad, or the State of Washington funding him. He wrote hopefully to the general agent for the Northern Pacific Railroad, asking for "the loan of a car for such a trip to advertise the state of Washington."[26]

On May 6 Meeker traveled from Seattle to Tacoma by steamer where he did some repairs to the wagon and picked up Dave and Dandy. He set up his typical camp on 12th Street and began selling his literature. The wagon was decorated with a twenty-eight-foot-long map of the Oregon Trail and Meeker's proposed national highway, "Pioneer Way." The map was courtesy of the Metropolitan Park Department.

On May 11 Meeker loaded the outfit on a steamer and traveled south to Olympia, where he spent a fruitless week attempting to secure funds for a monument near the site of Michael Simmons' 1850s sawmill on the Deschutes River in Tumwater. The rest of May was consumed in a very slow drive to Portland, slow because of Dandy's poor physical condition.

Oregon

Meeker arrived in Portland on June 2, 1914. He remained in that city through the first week of July, with a side trip to Vancouver for a lecture. He participated in Portland's annual Rose Parade and had a motion picture made of the event. On June 18 he spoke in the afternoon and evening to the Forty-Second Annual Gathering of the Oregon Pioneer Association. The evening talk included stereopticon views and motion pictures of his 1906–12 journeys east over the Oregon Trail. He also earned money by giving his standard lecture program at six different theaters around town. Unfortunately, that was when Dandy could go no further, and Meeker had him slaughtered. It was all business for Meeker. He expressed no regrets or sadness at the loss of his faithful ox.

Meeker's health was also problematic. He suffered an injury at the Rose Parade, healed slowly, and kept his daughter in the dark about it. In a long letter some months after the fact, he finally admitted to her that his ankle and foot had been caught between his cart and the ox Dave. "I thought it was crushed," he wrote, "but the people on each side of the street were clapping their hands and cheering and so I continued to salute them as we passed. As soon as the parade was over I got some [illegible] and bathed it but I tell you it hurt."

He continued, noting that a month after the injury, "My ancle is not well yet but is now better, thanks I believe to persistent bathing in hot water, massages and continuous use that has kept the blood from congesting; etc, etc, etc....Now what would have been the use to write this way and unsettle your peace of mind?"[27]

On July 10 Meeker started south, rather than east, as the dream of the eight-month advertising campaign ran into the reality that he was on his own and still broke. He would have to wait two more years to meet President Woodrow Wilson.

CALIFORNIA

En route to California Meeker filled forty-nine lecture engagements where he addressed 15,000 people and learned again the value of illustrations to accompany the spoken word. At these lectures he exhibited two short films of the ox team, two full films of Yellowstone Park, and a 1,200-foot film of the Rose Festival of Portland in which he appeared.[28] On August 10, 1914, Meeker, wagon, and the ox Dave arrived in Oakland, California, where he immediately began looking for other moving picture clips that he could include in his various programs. By the end of the year he added to his collection a short reel of moving picture scenes taken in Tacoma.

Meeker obtained a permit from the mayor of Oakland to sell his literature on the city streets, a total reversal of the difficulties he faced there in 1909. With that accomplished, he hired an assistant to look after the wagon and Dave and settled into the Clay-Ten Hotel. Ezra described it as a new, seven-story brick building with a hundred rooms. His room was "spick & span," with hot and cold water, a telephone, plenty of soap and towels, two windows, and plenty of sunshine in the early morning.[29] As he had no books with him as yet to sell, he began to search for opportunities to lecture. On August 26 he presented a program at a Unitarian Church and shortly after at College Hall in nearby Berkeley, and at several local theaters.

San Francisco, however, proved to be more difficult. His attempt to meet the mayor, George Ralph, was met with a rebuff. "I was denied access to the Mayor, the outer guard evidently looking upon me as not worthy of notice." He then asked Washington's governor for a letter of introduction. "I am striving to interest the people of this state to join us in the effort to get national aid for the memorial road through the South Pass (The Oregon Trail) in which the people of this state should be interested."[30] The governor was prompt in supplying the letter,[31] but by mid-November Meeker was still waiting to meet Mayor Ralph.

He told Curt Teich & Company, "I have been woefully disappointed in the results of my trip here and absolutely have not been able to send you the balance due."[32] He wrote his daughter that despite his troubles he had funds enough on hand to last two months even if he made no sales. On November 16 Meeker finally met Mayor Ralph who gave him a letter of introduction to California's U.S. Senator-elect James Phelan instead of a permit. While that might prove useful at a later date, it did not solve Meeker's immediate problem. No permit was forthcoming.

Not Wanted?

Meeker went to California with a gentleman's understanding arranged by the Metropolitan Park Board that the ox team and wagon would be exhibited in the Washington State Building at the Panama-Pacific Exposition and that he would be in charge of the exhibit. When he arrived he found that this might not be the case. He wrote Richard Seeley Jones, the Washington State Executive Commissioner. "As there has been no definite agreement as to myself, I write to ask what your understanding is, what compensation you think should be paid me and in what particular part you have in mind for me. I will not go in as a supernumerary; that is I am able and willing to render services for value received."[33] He wrote the Park Board, "In the event the commission refuses to carry out the agreement, I want to exhibit them elsewhere."[34] Meeker had been approached by another group participating in the Exposition, the California 49ers, who asked Meeker if he would set up his exhibit in their camp and promised him a salary of one hundred dollars a month ($2,387 in today's currency) and the right to sell his literature. Before turning down this lucrative offer he wanted to be certain he had a paid position with the Washington State exhibit. He wrote Jones again asking for an answer,

> [I]f it is not your wish to assign me a place I would ask you to frankly say so as I would not want to be a party 'imposed' upon the commissioners. If however it is your deliberate judgment that younger blood should be employed and that I should be sent to the 'scrap heap' so far as the state work goes, so be it, but you will realize the appropriateness of asking for an early answer.[35]

As it turned out, Jones wanted the now-stuffed ox team and wagon, but not Meeker. As the outfit legally belonged to the Metropolitan Park

Board, Ezra asked them what they thought. They stood firmly in Meeker's corner stating that Ezra was to "marshal this exhibit and be part of it. That was the understanding and agreement between this Board and the members of the Washington Exposition Commission." The board further noted that the agreement firmly stated that the Washington commission pledged to pay all expenses associated with the exhibit.[36]

By the end of November Richard Seeley Jones gave in. Meeker was offered a salary and was placed in charge of the ox team and wagon exhibit. Ezra wrote, "Mr. Seeley Jones has indicated he wished me to join the working force in the Washington State building at the exposition, to lecture. If confirmed by the commissioners it will incidentally afford a splendid opportunity to spread the gospel for a memorial road, based as well upon commercial importance."[37]

On December 4 Meeker brought the wagon to the exhibition gates preparatory to installing it in the Washington State Building. As he would not go on the payroll until February 1 he decided to make a quick trip to Seattle.

While these various dramas played themselves out, Meeker, ever the multi-tasker, continued to lobby for his Oregon Trail work. When he reached Olympia he contacted Governor Ernest Lister and updated him on the progress in the U.S. Congress, and asked if the governor would support a memorial to Congress requesting a survey of the Oregon Trail as a possible route for a transcontinental highway. Meeker told the governor he was working with the states involved to get them to do the same.[38] He then spoke to the Tacoma Commercial Club on the national highway question, worked with the state legislatures of Wyoming, Oregon, and Washington on the memorial, and celebrated his eighty-fourth birthday with his "old time friends" and some of his thirty-eight descendants.[39] At the end of December he told Seeley Jones he would ship the oxen shortly and would follow with the rest of the exhibit about January 20.

Meeker spent 1915 at the exposition, lecturing in the Washington State building several times a day, battling, as always, for his vision as to how things should be done. He brought his widowed daughter-in-law, Clara Meeker, and his granddaughter, Bertha Templeton, to San Francisco and found them employment at the exposition. All the while he continued to sell his books and postcards and promote the Oregon Trail bill in Congress.

Against this background, as World War I raged in Europe, Meeker saw a hopeful path to convince Congress of the necessity for his transcontinental highway.

PRESERVED BEHIND GLASS

It wasn't until January 2, 1915, that the Park Board officially accepted Meeker's gifts of oxen and wagon, and by then plans were in motion to transfer them to the Washington State Historical Society. In early July 1915 Secretary Bonney informed Meeker that the outfit would be stored after the exposition in the Ferry Museum at the request of the Park Board. Meeker then contacted the Smithsonian and asked if they still wanted the oxen and wagon. The answer was no. Many letters went back and forth between Meeker, Bonney, and the Park Board. In the end all concerns were worked out, and by January 1916 the oxen and wagon were in the historical society's museum. A floor-to-ceiling glass case to house the exhibit was completed in early 1917 at a cost of $400. On January 21, 1919, Meeker formally presented his papers, the wagon, and the oxen to the Washington State Historical Society. Meeker gave an address that day which in part included:

> I likewise assign, give and transfer to you the contents of the several boxes, now stored in the wagon containing the accumulation of many years of my correspondence; copies of letters sent, originals received, unpublished manuscript, moving picture films of my trip, copies of books published by me, in a word a brief, so to speak, of my life. There are letters in this collection, I do not wish to be made public during my lifetime, hence I will ask you to put the same under seal when I am called and so keep it until the first day of October 1952, being the date of one hundred years after my arrival in the Oregon country, a part of which is now known as the State of Washington.[40]

A notarized bill of sale was attached to Meeker's address. Both stipulated that the letters and papers were to remain sealed until October 1, 1952. Over the next fifteen or so years both Meeker and Bonney added documents to the collection in the wagon, but after Meeker's death, the historical society somehow forgot about them. Oxen and wagon remained sealed in the glass case until February 1963 when the curious museum director opened the case, looked in the wagon, and rediscovered the papers. In the next few years they were cataloged and placed in archival boxes.

The protective glass case was permanently disassembled in 1987–88 and the oxen and wagon taken off display until 1991 when they were installed in an exhibit space on the Ferry Museum's third floor. In 1995 this facility closed, reopening in 1996 as the Washington State Historical Society Research Center. Dave and Dandy and the wagon were moved to the new Washington State History Museum in Tacoma. At a later date the wagon was removed and stored in the basement of the Research Center. It was replaced at the museum with a facsimile wagon and a mannequin driver, but it is occasionally brought out for special exhibits. A somewhat dusty Dave and Dandy still stand at their posts, patiently greeting visitors.

CHAPTER 10

The Pathfinder Expedition, Part 1: 1915–1916

Beginnings

The idea for the Pathfinder Expedition was born at the Panama-Pacific Exposition where Meeker was lecturing under contract in the Washington State Building. His hopes to convince Congress to build a national highway across the continent following the route of the Oregon Trail had yet to be realized. He had been advocating this idea since his 1906–08 Old Oregon Trail Monument Expedition. By the fall of 1915 he had accumulated a war chest large enough to allow him to begin yet another campaign. And by that fall, as World War I engulfed Europe, Meeker saw the need for the United States to prepare itself for possible involvement in the conflict. He noted that during the Battle of the Marne, France stopped the German advance by rushing troops to the front using hundreds of automobiles over that country's "good roads." He felt a military road connecting the east and west coasts of the country was desperately needed. But what route should such a road follow? Obviously, the track of the Oregon Trail.

Automobiles had become a fixture in American life by 1915. Good roads had not. Meeker was well acquainted with both. Six years earlier one of the feature events at the Alaska-Yukon-Pacific Exposition (AYPE), held in Seattle, was a cross-country automobile race—the Guggenheim trophy race. Meeker hosted the awards dinner in his Pioneers Restaurant and heard tales of how difficult it was to cross the country by automobile. Nevertheless, the automobile industry was in full swing at the San Francisco exposition. Many manufacturers took the opportunity to display their products. Studebaker, Ford, and Pathfinder, among others, showed off their newest lines of automobiles.

119

Henry Ford even built a production factory on the exposition grounds that turned out multiple new automobiles daily.

Meeker posed for a photograph sitting in a Pathfinder motorcar at the exposition, and an idea was born. Why not build an automobile and make it look like an old time prairie schooner; equip it with a bed and a stove to make it a home on wheels—the nation's first motorized recreational vehicle (RV)? Then take it on the road to advertise the need to both save the Oregon Trail and build an efficient coast-to-coast military road. Experience demonstrated that crowds flocked to the sight of a covered wagon. Surely they would also flock to an automobile that looked like a covered wagon.

On October 4, 1915, Meeker wrote the Studebaker Corporation and inquired whether they would be interested in such a project. It would be an advertising bonanza for the company, he told them. He was politely rebuffed. He wrote again asking what it would cost to build such an automobile and received a discouraging reply. Studebaker told him the chassis alone would cost in the neighborhood of $1,200 and that placing a wooden wagon box with a bonnet on a metal chassis would not work, that it "would soon tear itself to pieces if adapted to a motor truck."[1]

Refusing to give up, Meeker wrote the company telling them that he intended to stop at their South Bend, Indiana, plant on his way to Washington, DC, at the close of the exposition to discuss his idea further. Ezra explained that what he envisioned was not simply placing a wooden wagon box on a chassis, but something that looked like an old-style prairie schooner. He suggested calling the vehicle "Schoonermobile or "Prairie Schoonermobile."[2]

At the end of November Meeker was developing a contingency plan in the event that Studebaker continued to rebuff him. On November 23 he floated the idea with the Pathfinder Company in Indianapolis. He mentioned that he was often called the "Pathfinder" and sent them some of his Oregon Trail literature. "I will leave here for Washington City via New York on the first of December. If you thought it worthwhile I could arrange to stop off at Indianapolis."

He received a positive reply and made arrangements to stop in Indianapolis on his way east. However, Meeker had not yet given up on Studebaker. He bypassed the Studebaker advertising department and contacted eighty-two-year-old company founder John Mohler

Studebaker and offered to give a lecture on the Panama-Pacific Exposition and the Oregon Trail during his stop at South Bend. Mr. Studebaker was interested and sent Meeker some sketches of a potential Schoonermobile. But he made no promises. "I am a vehicle man, and if you come to South Bend I will turn you over to the automobile end of the business, and if you can show them that there is something in it for the Studebaker auto, why you might interest them."³

Meeker departed San Francisco on December 1 and made his way to South Bend. After meeting with John Mohler Studebaker, he spent three days unsuccessfully lobbying the company to support his plan. A week later Meeker was in Indianapolis at the Pathfinder Company where he met with a Mr. F. G. Buskirk, the assistant sales manager, with whom he discussed the proposed trip over the old trail. After consulting with the board of directors, Mr. Buskirk told Meeker that the company would loan him a car in which to make the trip across the continent for the publicity the company would receive.

Meeker then continued on to the nation's capital where he celebrated his eighty-fifth birthday by buying a new set of clothes.⁴ He spent the month of January in his new suit lobbying Congress to pass his national road bill.

Before he left San Francisco Meeker had secured a tentative business agreement with N. E. Sheffield and his wife, a contact he may have made at the Panama-Pacific Exposition. If Meeker succeeded in obtaining his traveling car, Sheffield was to drive it across the country. His wife would serve as Ezra's secretary taking dictation, typing letters, and assisting with his newest book, *The Busy Life of Ezra Meeker's Eighty-Five Years*. Meeker planned to print 5,000 copies of the new book in Indianapolis and 20,000 copies of his pamphlet, "Story of the Lost Trail to Oregon, No. 1." He had $1,000 on hand—not enough to print his book and buy an automobile, let alone a customized one. He needed an arrangement whereby the car would be provided gratis.

He sent for the Sheffields, and on February 11 they arrived in Washington, DC, prepared to take up their duties. That day the Pathfinder Company wrote Ezra expressing their concern about his choice of an inexperienced driver. A week later a second letter arrived, again stating concerns about Sheffield, and telling Meeker that the automobile would be ready within sixty to ninety days. Ezra informed the company that he had the utmost confidence in Sheffield and told them he wanted the deal finalized before he advertised the car in his book.[5] There are several photographs of Meeker sitting behind the wheels of various automobiles, giving the impression that he knew how to drive. However, he admitted in his unpublished story about the Pathfinder Expedition that he did not have that skill.[6]

President Number Three

On February 13 Meeker met with President Woodrow Wilson, an audience arranged by the Washington State congressional delegation.[7] Meeker asked the president to support the Pioneer Way bill working its way through the Congress. With unprecedented foresight Meeker told Wilson:

> Mr. President; I lived on the Pacific coast when Commodore Perry sailed into the ports of Japan; lived among the Japanese from the time they began to arrive on the coast to the present day. The Japanese are a proud people. They believe they are as good as any other people, and a little better. History will repeat itself. There will come a time when there will be war on the Pacific coast. Do you remember how completely our railroads failed in the transportation of troops and munitions in the time of the Spanish war?... This proposed Highway is a great measure of preparedness and it would seem there does not live the man with any red American blood in his veins that would not favor such preparedness. The motorcar is now the great factor in movement of troops in the wars of Europe, but to utilize this factor we must have the road bed, a thoroughfare with easiest curves and grades obtainable.[8]

In a letter to his daughter, Meeker described his meeting with President Wilson, and his reaction to the bill,

> He said he was in favor of the proposition. I had heard he had a way of listening to people saying but little himself; but he did not treat me so, but seemed to enter into the conversation with animation and deep interest and frankness. The interview was short but very satisfactory and I believe we have a friend for the bill in the White House and [am] encouraged to believe it will pass.[9]

Meeker told William Bonney, Secretary of the Washington State Historical Society, "He is ours... The President treated me very cordially and without any reservation said he would favor the movement and I think he will so communicate to the Congress.... I am promised a hearing before the Military Committee on Friday of this week."

After testifying before the committee, Meeker wrote Bonney, "[I] am egotistic enough to believe I made a favorable impression and expect to get a favorable report in a few days."[10]

NEGOTIATIONS

Meanwhile, negotiations with the Pathfinder Company about a driver were moving at too slow a pace to suit Meeker. In mid-February he wrote William Bonney reciting his troubles with the company and asking for a loan to purchase the car, if needed. Bonney, like many of Meeker's family, friends, and acquaintances, was used to ignoring these requests, and did.

Meanwhile, the Sheffields were putting Ezra's new book in order. They worked a split shift—8 a.m. to noon and 6 p.m. to 10 p.m. This gave them the opportunity to spend the afternoon hours seeing the sights of the nation's capital.

In early March Meeker wrote the Pathfinder Company and threatened to call the deal off over the driver dispute. Apparently they compromised, and Mr. Sheffield was sent to Indianapolis at the end of March to learn firsthand from the company how to drive their car. Carrie Osborne wrote her father in early April, "Here's hoping Mr. Sheffield gets the making of the machine so thoroughly that you will have no dismal breakdowns. Tell him not to let the schooner mobile break only in good paying communities. You will have to find an elevated name for that car."[11] In fact, the Schoonermobile was depicted as such on the cover of *The Busy Life of Eighty Five Years*, but the name disappeared from Meeker's writing almost immediately. He and everyone else referred to the vehicle as simply the Pathfinder.

By mid-April the Pathfinder was ready and Meeker journeyed to Indianapolis. The car more than met his expectations, and his new book had also arrived from the printer. He was finally set to go, but with one problem. For reasons unknown, likely some unresolved friction, the Sheffields severed ties with Meeker. Whatever the reason, the Pathfinder Company expressed their concern. "We trust that this personal difference between you and Mr. Sheffield will not in any way interfere with your ultimate intention of making this transcontinental trip…this Company has a big interest in your trip and we would hate to have anything come up of unpleasant nature which would become public property and detract from the dignity of your mission."[12]

Pathfinder in Indianapolis. Note the driver behind Meeker. *Author's collection*

On the Road

On April 22, 1916, with Earl S. Shaffer at the wheel (a driver possibly recommended by the company at short notice but whose salary was paid by Ezra), the Pathfinder started on a seven hundred mile trip east to Washington, DC, from Indianapolis. The vehicle weighed 4,450 pounds empty and 5,165 pounds loaded. It guzzled gas at the rate of seven to eight miles to the gallon. On the bonnet was an eighteen and a half by forty-eight-inch map of the Oregon Trail. The odometer had eleven miles on it when Meeker hit the road. On that first day the nation's first RV traveled sixty-nine miles. Meeker and Shaffer camped by the roadside three miles east of Richmond, Indiana. It took six days to traverse 710 miles. At 9:30 a.m. on April 27 the Pathfinder rolled into Washington, DC. Meeker recorded in his day calendar $1.80 in incidental sales along the way, having had no time to make sales stops. The following week was spent in the capital selling literature from the Pathfinder. The official start west began at the White House in the afternoon of May 5. Newspapers gave Meeker a well-publicized send-off.[13] Six days later the Pathfinder arrived in Pittsburgh for a three-day stay. Meeker wrote the company and informed them that sales to this point were a disappointing $122.55. They responded with a note of regret "that the financial end of your trip is not a success," and then brought up a subject that appeared often in the correspondence. "We note in the write-ups you occasionally leave out the word "Pathfinder" and we would like for you to handle this matter in such a way that the publicity secured will show that it is a Pathfinder Twelve cylinder car being used for the trip."[14]

The Pathfinder continued on through Wheeling, Zanesville, Columbus, and Dayton. Here Meeker noted that sales so far totaled $317.64 and that 74 gallons of gasoline had been purchased. All was not well with Mr. Shaffer. Upon arrival in Indianapolis Meeker and Shaffer parted company with some acrimony. The driver complained to others that Meeker had not settled with him according to their contract. Meeker wrote Shaffer, "You will remember that I paid you to the last farthing as rendered in writing by you at our settlement in your bill made out at my request." Meeker went on to remind his now ex-driver that he paid his fine for driving without a license in Maryland, keeping him out of jail. He ended by stating, "I will give you a piece of advice, that if you will follow, will be of great value to you through life. If you have aught against your brother, go straight-way to Him, and not to others."[15]

Meeker remained in Indianapolis ten busy days. He made arrangements to store five boxes of books and a trunk with the publisher of his new book; sent $55.82 to Mr. Sheffield and noted there was a balance of $25 still due him; and hired John Mattlock to drive the Pathfinder to Chicago. As always, Meeker kept in close touch with his allies in Congress. At his stop in Indianapolis he visited the father of Congressman William Humphrey, and Humphrey wrote Ezra that he hoped the two "boys" stayed out of trouble.[16] On May 30 "Jack" Mattlock and Ezra started for Chicago.

While in the Windy City Ezra received a letter from William Mardon with a clipping from *New York Mail* about the cross-country trip.[17] Before leaving Chicago Meeker hired his fourth driver, Charles J. Wyley, age twenty-four, of Oak Park, Illinois. The first stop was Elgin, Illinois, followed by an excursion north to Clinton and Beloit, Wisconsin. Then back to Illinois, stopping at Rockford, where they camped in the city park. All along the route Meeker sold copies of his newest book, *The Busy Life*, and other Oregon Trail materials. Ezra also kept the Pathfinder Company somewhat informed of his troubles and successes and described their difficulties in getting mired in the muddy roads. The company wondered if the driver was being careful enough. "An automobile has certain limitations; one of them is… that they lose traction in extremely bad road conditions or mud, where a horse and wagon can go right thru, a machine becomes stalled. An experienced road driver should be able... to determine by examining his road conditions whether the car will go thru or not. We suggest your cautioning him along these lines."[18]

At the end of the month Meeker also received a not so subtle reminder of the business arrangement he made with the Pathfinder Company. Wrote F. G. Buskirk, "We are going to ask you, again, when talking with the newspaper men in the various towns you... impress upon them the fact that you are traveling in a Pathfinder Twelve Cylinder Car. We thought this was thoroughly understood when you were here."[19]

On June 30 in Peoria, Illinois, a new driver appeared without explanation in Meeker's day calendar, Ernest Grot, a student at the University of Illinois. Grot would stay with Meeker all the way to the West Coast before returning to the university in the fall. Apparently Grot was suggested by the Pathfinder Company, which felt that the vast majority of the trouble Meeker was having with the roads was "due to an inexperienced driver with a twelve-cylinder motor."[20]

As he worked his way through Illinois, Meeker, heeding the previous admonitions, supplied the Pathfinder Company with a weekly itinerary and in turn they sent press releases, photographs, and advertising material to the cities along his route. The company was pleased to note that Ezra was now highlighting the automobile in his interviews with the press. "We are expecting some big publicity from St. Louis and be sure to have them bring out Pathfinder as much as possible. This is a big town and we should get some wonderful publicity out of it."[21]

Meeker arrived in St. Louis on July 9. The next day he met with Walter B. Douglas of the Missouri Historical Society. What they discussed was not recorded, but no doubt it involved the Oregon Trail. In Jefferson City Ezra wrote William Bonney outlining how he had contacted and lobbied the congressmen of fourteen states, and informed him that his expenses were running about six dollars a day. While crossing the state the Pathfinder began to experience occasional mechanical problems. In response, the company advised Ezra to consult only "expert mechanics" and was pleased to hear the Grot was working out well as a driver.[22] Meeker must have been pleased as well for Ernest Grot stayed with Meeker all the way to San Francisco, and the revolving door of new drivers slowed for a time. On July 20 the expedition arrived in Kansas City. The odometer read 3,755 miles. Ezra Meeker was now poised to take his RV over the entire length of the Oregon Trail.

OVER THE OLD TRAIL ONCE AGAIN

While in Kansas City, Meeker received some good news from the Pathfinder Company, offering to assist with expenses in the form of a weekly check of fifteen dollars to pay the driver. "We believe with this assistance that the financial worries of the trip will be over and that you will be able to give it the attention you think it deserves."[23]

The Pathfinder needed some mechanical repairs before it could hit the trail. The Western Motor Company billed Meeker $2.50 for six feet of brake lining and ten cents for "one left washer."[24] Repairs complete, the odometer reading 3,850 miles, Meeker left Kansas City on July 28 and started west over the old trail.[25] The first day Grot and Meeker covered thirty-two miles to Gardner, Kansas, and then seventy-two miles farther to Lawrence, where they spent the night. The next stop was Topeka where the local newspaper reported, "Ezra Meeker, 85, who was retracing the Oregon Trail in an 80-horsepower Schoonermobile, arrived in Topeka. He was traveling from Washington, DC, to Olympia, Wash., to promote a national highway."[26] No doubt the Pathfinder advertising division was irritated at the lack of mention of the company name in the article. According to his day calendar Ezra met with William Connelly, Secretary of the Kansas Historical Society, while in Topeka. Again there is no record of their conversation, but as in Missouri, it was likely about the Oregon Trail and Pioneer Way.

From Grand Island, Nebraska, their route took them along the north side of the Platte River, following the path Meeker took in 1852. They arrived at Scottsbluff on August 9, with 4,117 miles on the odometer. Meeker was meticulous in keeping his accounts. Sales were good here—$21.65, but thirteen gallons of gasoline at twenty-five cents a gallon took $3.25 out of his wallet. He also received a complaint from the Pathfinder Company about not keeping them posted with his weekly itinerary.[27] An effort to get the company to increase the driver's weekly stipend was rebuffed. "We were willing to spend $15.00 a week to assist you in taking care of his expense. If there is any differences in the amount your driver expects and the $15.00, it will have to be taken care of by you."[28]

For some reason Meeker detoured north to Lusk, Wyoming, before moving on to Douglas and Casper, where the duo tarried for four days. Sales were a brisk $81.20. The Lander, Wyoming, newspaper of August 18 supplied information as to Meeker's activities in Casper.

> Ezra Meeker, who is known by millions of men, women and children of the United States in connection with the old Oregon Trail, was in Casper last week on one of his numerous trips across the United States, and is traveling wherever possible by the old trail, which while almost extinct in places, is plainly marked in others. At the present time 150 granite monuments mark the trail from Washington, D.C. to Puget Sound, many of the monuments in Wyoming being placed by Captain H. G. Nickerson of Lander. Mr. Meeker is not driving his prairie schooner this trip, but has instead an up to date, 12 cylinder Pathfinder truck. The octogenarian traveler was met at Casper by Governor Brooks and a number of citizens, and left for the west on Monday evening.[29]

Meeting with elected officials was always a Meeker priority. The old pioneer called upon President Wilson, U.S. congressmen, governors, state legislators, and mayors of cities on his way across the country on behalf of *H.R. 9157, Pioneer Way*. Everywhere he preached the need for a permanent road that could be used as a military highway.

Meeker purchased eighteen gallons of gasoline before departing Casper and on August 15 headed for Independence Rock. At Devil's Gate they pulled into a courtyard with vacant buildings on one side and a barn on the other to escape the fury of a dust storm. While looking around for shelter, Meeker noticed an old-time stagecoach and, with a finger in every pie, had two hours later located the owner, Charles Fletcher, and secured it for the Washington State Historical Society. He made arrangements for shipping it to Tacoma and wrote that Bonney intended to place it in the museum beside Dave and Dandy.[30]

The Lander newspaper reported the Pathfinder's arrival in their town, reporting that Meeker was the guest of Captain Nickerson, president of the Wyoming Oregon Trail Commission throughout its existence from 1913 to 1921.[31] The story noted that Meeker had been in Lander ten years prior with his team of oxen.[32]

Two days after leaving Lander the Pathfinder arrived at the South Pass Oregon Trail summit monument that Meeker placed in 1906. Here he paused and took some photographs. The next day the Pathfinder passed through Cokeville where another photograph was taken. On August 20 the expedition stopped at Montpelier and Meeker took a photograph of the car next to that city's ten-year-old monument. He took a photograph a few miles up the road at Lava Hot Springs as well. All of these photographs appeared in Meeker's booklet, *Story of the Lost Trail to Oregon, No. 2*, the only place they seem to have been published.[33]

Yellowstone Sojourn/ Success at Fort Hall

Upon arriving in Pocatello, Idaho, Meeker visited Mrs. Emma Standrod, founder and regent of the Wyeth Chapter of the Daughters of the American Revolution,[34] most likely to discuss upcoming plans for another attempt to locate the site of the original Fort Hall. But first Meeker decided to make a slight detour. Neither he nor his driver had seen Yellowstone National Park, and here was their opportunity. They loaded up with twelve gallons of gasoline and drove from St. Anthony, Idaho, to the west gate of the park. The tourists spent four days visiting such sights as Yellowstone Falls, Dragon's Mouth, Keppler Cascades, Lower Geyser Basin, Mammoth Paint Pot, Artist Point, and the Junction of the Madison and Gibbon Rivers. Meeker, ever the penny-pincher, let Grot know that this holiday would be off the clock, with no pay forthcoming. No objection from Ernest was noted, however. Photographs of the visit were also included in the pamphlet *Story of the Lost Trail to Oregon, No. 2*, which undoubtedly pleased the Pathfinder advertising department.

Back in Pocatello, Meeker joined Minnie Howard,[35] a medical doctor and local historian who later became Meeker's partner in founding the Old Oregon Trail Memorial Association (OTMA), in an automobile entourage that went searching once again for the original Fort Hall site. Meeker considered this the most important historic point on the Oregon Trail, and it was his third try at locating the elusive site. In May 1906, when Meeker stopped at Pocatello on his first Monument Expedition, a local man named George North took him out to what he thought was the Fort Hall site near the Spring Creek Bridge. After doing some excavations and coming up dry, Meeker concluded that Mr. North was in error in choosing that particular location. Even so, sometime after his 1906 visit a stone monument to mark the Fort Hall site was incorrectly placed there, where it remained until 1920 when it was removed to the true Fort Hall site.

Meeker's second attempt to find the site in 1912 resulted in the same negative outcome—no fort. He found the true site on his third attempt in 1916 with Dr. Minnie Howard, her husband Dr. William F. Howard, and Joe Rainey as guides. Excavations turned up numerous artifacts of the old fort. A temporary wooden post was set, subsequently replaced with a stone marker. The Fort Hall site played a prominent role in the course of events over the next decade. In 1921 Meeker returned to the site where he gave a speech in honor of Jason Lee's sermon, the first west of the Rocky Mountains. In 1925 efforts by Minnie Howard and others to memorialize the fort site became a stepping-stone to the birth of Meeker's Old Oregon Trail Memorial Association, a topic addressed in full in chapter 19.

Heading West

On August 30 the Pathfinder Expedition resumed its westward journey. Meeker stopped at American Falls and spent the night at nearby Burley, Idaho. From Twin Falls he sent a telegram to the B. F. Goodrich Company in Kansas City telling them he needed both new tires and inner tubes. He had tried Portland and Salt Lake to no avail. The Pathfinder Company was not too helpful in this search. They wrote Meeker that he could probably get them from St. Louis but suggested that he look around in Seattle when he arrived there. "We are glad to hear that you are going to stir things up when you get home," wrote the company's director of services, "and we hope that your speaking campaign will be the means of our selling a number of cars."[36]

On September 3 the Pathfinder reached Boise, Idaho's capital city. Here Ezra met Mayor Samuel Hays and Governor Moses Alexander. Meeker also urged the commercial club to get behind the Pioneer Way effort. He described Idaho's roads as both "rough" and "good." "We had to go slow in some places," he said, "but occasionally were able to move right along, as in the first 29 miles between Mountain-home and Boise, which we covered in 50 minutes." Meeker also mentioned that there were 160 markers along the trail.[37] Then it was on to Portland, Oregon, and Washington State. On September 14, 1916, Ezra and Ernest arrived in Seattle, mile 7,303. After a two-year absence Ezra Meeker was home.

The Pathfinder pulled to a stop in front of the *Post-Intelligencer* building in Seattle where Meeker was interviewed. He raved about the comfort of the automobile stating that he was less fatigued after a two-hundred-mile day from Vancouver than he was after a sixteen-mile day with an ox-team. He said, "I have never been up in an airship, but I believe the twelve cylinder automobile is the nearest approach to that sensation that a human being can have and still remain on the ground."[38]

After a brief stay in Seattle Meeker returned to Puyallup. His first stop was at Pioneer Park, the site of his old cabin. The Puyallup newspaper reported,

> The old prairie schooner body of the automobile attracted much attention wherever it stopped. It is covered with thousands of names and addresses of persons all along the route. Big signs on the car, telling of the purpose of the trip were almost obliterated. A big map of the route is also shown, the old Oregon Trail extending from Olympia to Kansas City, and the national old trails road from Kansas City to St. Louis, and thence to Washington, D.C. by the Cumberland road. It is 3,560 miles long.[39]

Next Meeker took the Pathfinder to the Puyallup Fairgrounds where he spent several days greeting many old friends and where the Pathfinder drew much attention.[40] On September 21 Meeker went east over the Cascade Mountains, stopping in Ellensburg,[41] then Yakima where he

Pathfinder, Ernest Grot, and Ezra Meeker in Seattle. *Author's collection*

attended the state fair. The governors of Oregon, Idaho, and Washington were there, as was William Bonney who was looking for some historical graves. Officers of the state Daughters and Sons of the American Revolution (DAR/SAR), involved in marking the Two Buttes battlefield in Union Gap, were also in attendance, and the Yakima Pioneers society was just being organized. It was the type of event that Meeker wouldn't miss.[42] As an aside, Meeker noted brisk sales of his literature.

A week later the Pathfinder was in Bremerton and Meeker was speaking at the naval shipyard gate urging the construction of Pioneer Way as a matter of military preparedness and better communication between the states. That city's newspaper succinctly summed up what Meeker had been doing for the past year and past decade: "four months in Washington working for the passage of the bill…providing for a survey for a military and post road from St. Louis to Olympia following the old pioneer trail. Monuments to the number of 160 have been erected along the way marking the old route to the west…largely through the efforts of Mr. Meeker directly or indirectly."[43]

On October 2 Ezra wrote the Pathfinder Company with a litany of mechanical problems. When they arrived in Seattle the automobile seemed to be in perfect running order, but as a precaution Mr. Grot thought they ought to put it in the shop and check everything out. The bill was $39.40. Then a spring broke and a gasket blew out on the way to Yakima. Next the frame was discovered broken about four inches from the back end of the front spring, necessitating some welding.[44]

Repairs completed, Meeker visited the new DAR/SAR Oregon Trail monuments between Tumwater and Centralia.[45] Shortly thereafter he received a letter of praise from the advertising department of the Pathfinder Company. "I want to congratulate you on the wonderful amount of publicity you have gotten. Really, Mr. Meeker, it is almost unbelievable. As near as I can figure it is in the neighborhood of four hundred full newspaper pages, and you know that's going some." Magazines such as *Colliers*, *Leslies*, *Modern Methods*, *Christian Herald*, and *Southern Woman's Magazine* all ran feature stories about the Pathfinder Expedition. Meeker saw nothing unusual in generating this kind of publicity. He had been doing it for a decade. The company also suggested that Meeker write up the story of the trip based on his *Lost Trail No. 2* booklet.[46]

With the Pathfinder repaired and his head full of praise Meeker scheduled a trip to Everett to speak about his adventurous trip across the continent. The *Everett Herald* reported on the reaction to Meeker and the Pathfinder: "Immediately after his arrival in Everett yesterday scores of curious persons surrounded the strange looking car and, armed with pencils, proceed to sign their names to all parts of the machine." This was not unique. His 1906 prairie schooner was covered with similar signatures and graffiti. The story pointed out that Meeker's Pioneer Way traversed 14 states and touched 107 counties having a population of 5,685,513 people who would be served by this national highway.[47]

In Everett, yet another problem surfaced with the weary Pathfinder. The generator was not working correctly, a problem that first occurred shortly after the Yellowstone Park detour. A local electric works attempted a repair that lasted about three minutes. Meeker next took the generator to a Westinghouse service department where the repair worked for twenty miles. He then requested the company send him a new generator. He also informed the company of his latest scheme to extend his travels south through Oregon and California to San Diego.[48] Meeker and the company then worked out the details of this proposed trip. The Pathfinder Company agreed to continue the subsidy for Ernest Grot as long as Meeker kept giving them advance itineraries.[49]

The next political battle lay ahead, the wooing of California's congressional delegation. The concern was that they were inclined to support funding for the Lincoln Highway, conceived in 1912, that was to run from New York City to San Francisco, rather than Meeker's Pioneer Way. Ezra laid out his thoughts in the *Tacoma Ledger* on how to accomplish both goals. He didn't see the two as incompatible.

> It is my firm belief this opposition if it exists, can be disarmed and turned into active support by the California delegation. I remember that in 1852 we that were on the way to Oregon were mingled with throngs of people on the way to California and for more than 1,000 miles, and that at parting on Bear River San Francisco was nearer than Portland. Pioneer Way will serve as a trunk line through the south pass of the Rocky Mountains for the Pacific slope, California, as well as Oregon and Washington. The Lincoln Highway will serve the people of California by connecting with Pioneer Way near the old time parting of the way.[50]

The odometer stood at 8,262 miles when Meeker and Grot departed from Tacoma.

CHAPTER 11

The Pathfinder Expedition, Part 2: 1916

Seattle to California

On October 9, 1916, Meeker wrote to Charles Davis, founder of the National Highways Association. The purpose of the organization, incorporated in 1912, was to promote a national highway system. Its first effort was to publish a map of a proposed 50,000-mile system and distribute it around the country. In 1916 nearly forty people were employed at the association office in South Yarmouth, Massachusetts, drafting state highway maps, bulletins, circulars, and pamphlets to the tune of some fifteen million copies, all aimed at getting Congress to create the national system. Meeker, while always fixed on his own national highway idea, would work closely with Davis and the organization for over a decade. In the 1920s Meeker moved to New York City and worked out of the association's New York City office. From there he would found the Oregon Trail Memorial Association in 1926. In his letter to Davis, Meeker spelled out the situation with Congress and the prospects of getting his Oregon Trail survey bill passed. He told the National Highways Association leader that the bill would pass in the Senate, under the leadership of Senators William Stone of Missouri, George Chamberlain of Oregon, and Wesley Jones of Washington. He was certain that the twenty-eight senators from the fourteen states the route went through were a lock.¹ He told Davis the House would be a more difficult "nut to crack," and asked for the association's support.

> In my recent trip over the proposed route I interviewed a large number of influential people, Governors of states, mayors of cities, commercial and automobile clubs asking them to express their influence upon their Congressmen to actively support the bill. With but one exception they promised to do, but realizing the forgetfulness in such matters I want

to take this matter up by mail send them a copy of the map, write a personal letter as likewise to many Congressmen with whom I have a casual acquaintance or whom I think will be interested. You realize what labor will be needed to prosecute such a campaign and that it won't be physically possible to do this without a stenographer to assist.

In any event I want to visit California. The impression is abroad that California will oppose the measure to survey Pioneer Way. My belief is I can disarm any thought of opposition and in fact believe we can secure their active support.[2]

The Pathfinder Expedition, Part 2, was about to get underway with Meeker and his driver, Ernest Grot, heading south from Tacoma through western Washington. In Centralia he gave a short talk to a crowd on Tower Avenue advocating Pioneer Way and also giving a plug for presidential candidate Charles Hughes.[3] The *Chehalis Bee Nugget* gave his effort a boost, commenting that he had made four trips across the continent, 1852, 1906, 1910, and 1916 and was now embarking on his fifth such trip, albeit via California.[4]

On October 15 the Pathfinder arrived in Portland, Oregon. While there, Meeker met with city officials and representatives of commercial and automobile clubs in an effort to get their active support in backing the plan for Pioneer Way. At these meetings he told the usual stories of his efforts to get the bill through Congress, tales of the Pathfinder trip across the continent, and touted the value of the proposed road as both a military highway and as a memorial to the pioneers. He also advocated the commercial value of such a road.[5] Case in point: there were no decent roads over the Siskiyou Mountains of southern Oregon and northern California. Meeker gave some thought to bypassing this route in favor of going east of the Cascade Mountains to enter California over Donner Pass, but he changed his mind.[6] He notified the Pathfinder Company that he planned to travel south to San Diego, then take a southern route from there to Washington, DC. If that route was found practicable, he would arrive in early January. He said he had the financial resources to see his way through. He thought the Pathfinder was in perfect working order and he wanted to get over the Siskiyous before the fall rains arrived and hoped to be in Sacramento the last week of October. He promised to send an itinerary as soon as possible.[7]

When Meeker and Grot left Portland on October 19, the odometer read 8,472 miles. Two days later they were at Medford. Problems with the Pathfinder arose almost immediately. First the generator; Meeker requested a new one be sent to San Francisco. Then the splash plate in the gasoline tank broke loose and disabled the fuel gauge. Next up were the spark plugs that fouled every hundred miles and had to be removed and cleaned. Grot thought the problem was loose pistons. They asked the company for advice.

On October 22 the expedition crossed the California state line. In Hornbrook, California, Meeker recorded in his day calendar that they broke a spring. In Dunsmuir he complained of the bad road. Two days later in Redding he received a letter from the company telling them that a new generator was on the way, but they were on their own with the other problems. The Pathfinder passed through Red Bluff and Oroville. In the latter city Meeker again strayed into presidential politics and made an appeal to the populace to vote for Hughes in the upcoming election. He also wrote the Pathfinder Company telling them he had repaired the broken leaf spring. "It's a mighty car when in good humor but has taken a lot of petting to keep it so."[8]

To Charles Davis he wrote, "My mission here is to disarm any opposition from California; in fact I hope to enlist the active support of the California delegation in Congress. To do this I must convince them it's to California's interest to do this." He went on to request that the association print copies of a map of the trail between Soda Springs and California to show the California delegation how much they shared the route of Pioneer Way with the northwest immigrants. He advised Davis that he intended to "make as strenuous campaign I know how to do and not make a nuisance of myself."[9]

On October 31, 1916, Meeker and Grot arrived in San Francisco. The Pathfinder covered the 136 miles between Sacramento to San Francisco via Stockton in five hours. The car went into a local repair shop and Meeker checked into a hotel. He remained there nearly three weeks. The new generator arrived and was installed, but the battery would not hold a charge. Even so, it was deemed good enough by a local mechanic to get him to Los Angeles. As was often the case in his visits to San Francisco, sales of his literature were weak, so Meeker entered the Pathfinder in a Good Roads Parade in hopes of generating some publicity. Much of his time was spent lobbying California congressmen.[10] He also began soliciting opinions, somewhat late in the game, as to whether it was feasible to drive the

4,400 pound vehicle (5,200 pounds loaded) across the southern tier of the United States.[11]

When Meeker received the maps he had requested from the National Highways Association he wrote Davis, "I find California a harder nut to crack than I had anticipated to line up for Pioneer Way—think however they will come over after election."[12] He also requested financial help from the association for the final push to Washington, DC. From his daughter he received election news. Washington voted in prohibition, going "dry" by 100,000 votes. Wilson defeated Hughes in the presidential contest.[13]

The local automobile club told Meeker that it would be almost impossible to get the *San Francisco Examiner* newspaper to mention the Pathfinder or Meeker's mission and declined to give him a letter of introduction to the editor. They misjudged. Meeker called on the editor and invited him to send a photographer in the car to the Oakland Hills, where he intended to pay his respects to the memory of California poet Joaquin Miller. He told the editor that he would write an article about the visit for the newspaper, or if they wished they could send along a reporter. To gather his own photographs of his visit Meeker engaged a second photographer to go along. Also, the Oakland Chamber of Commerce held an automobile "review," the purpose of which was to solicit support for the Lincoln Highway. They invited Meeker and his Pathfinder to participate. He did, decorating the vehicle with a large Pioneer Way map. Some 20,000 viewed the Pathfinder and were exposed to the idea of Pioneer Way. "Five moving picture cameras took the car with previously arranged programs and I don't know how many map shots, but a good many."[14]

With the departure of Ernest Grot in San Francisco, presumably to return to school, the immediate question facing Meeker was who would drive the car to Los Angeles. He appealed to his sixty-four-year-old son Marion, who lived in San Bernardino, California, and sent him money enough to take a train to San Francisco. Marion responded, "Have plenty of house room, and good place for your Pathfinder. Yes I drove a car, and am said to be a good careful driver. I prefer not over 25 miles speed and usually run from 20 to 22 or 23. However can make 35 or 40 but don't like it. Slower is safer and 25 is the limit under the state law."[15]

Marion arrived via the Santa Fe Railroad on November 17. Father and son shortly after took their leave from San Francisco. On the afternoon of November 24 they "limped into Los Angeles." The Pathfinder once again headed to the repair shop, Ezra took a room at the Hotel Rosslyn, and Marion went home for a time. Meeker would stay in southern California through Christmas. Ahead was the trip east via the "southern route" and more adventures than even Ezra Meeker could imagine.

Southern California

On December 2 Meeker made his way to San Bernardino, leaving the Pathfinder in the repair shop.[16] The local pioneer society had invited him to speak about his project and his travels and he eagerly accepted.[17] Meeker was joined on the stage by another pioneer, William S. Tittle. Both men were born in 1830 and both came west in 1852. The difference was their method of travel. Tittle came by ship around Cape Horn. Meeker came by ox-team. The local newspaper reported that Ezra captivated his audience. "The hearers were held in rapture of the past as Brother Meeker told of his life from a small boy…to his recent trip back to Washington D.C. in an 80 horse power Pathfinder schoonermobile."[18] During the gathering Ezra, son Marion, and granddaughter Grace Meeker were inducted into the pioneer association and arrangements were made for Meeker to show one of his motion pictures during his stay. Then the chairs were cleared away and the customary quadrille commenced, led by Ezra Meeker, age eighty-five, and Mary Crandall, age eighty-two.

Four days later, with Marion still driving, Meeker was in San Diego with the Pathfinder, where he gave a lecture and met with the mayor, the Chamber of Commerce, and the head of a local exposition to enlist their aid. While in San Diego he advocated a national highway along the Mexican border in addition to his Pioneer Way. "I believe we should have both of them as a military and preparedness measure…. I am attached to the old Oregon Trail both for practical and sentimental reasons, but a road along the border is as much a necessity as is a highway across the continent farther north."[19]

On December 12 Meeker was in Long Beach. The local newspaper informed their readers that "Ezra Meeker, Indian fighter, pioneer settler and advocate of military preparedness for the United States as a preventative measure for war, arrived in Long Beach this morning with his new 'prairie schooner' after having covered a distance of nearly 11,000 miles in seven months." Meeker must have been surprised to learn of his career as an Indian fighter. Still somewhat challenged in reporting the actual facts, the newspaper dubbed Ezra the "official representative of the National Highways Association."[20]

With these credentials duly noted, Meeker visited the Chamber of Commerce and went before the city council where he gave his customary address. A large crowd, full of questions, gathered around the Pathfinder when it arrived at the chamber headquarters. Meeker told his audience that he had basically lived in the car since he started west from Washington, DC, on May 5 and that it was much more practical and useful than the old style prairie schooner.

On December 15 Meeker invited the San Bernardino Chamber of Commerce to support his effort. Marion Meeker also addressed the chamber and outlined the future value of his father's "Pioneer Way" to California. Though expressing much interest, the businessmen of San Bernardino decided they needed to study Meeker's bill before endorsing it.[21] Nearby Redlands also received a visit from the Pathfinder. The newspaper reminded its readers of Meeker's last visit to their city in 1910 with an ox-team. It reported Meeker "unchanged and enthusiastic as ever," and "while Mr. Meeker claims to miss his oxen he says he is more than compensated by the added comfort and speed."[22]

Daughter Carrie Osborne was surprised to learn that Ezra was about to embark on a drive across the southwestern deserts. She sent him a photograph and a few pairs of warm socks for a combined Christmas and birthday present and gave him the bad news that his friend and frequent financial supporter Louise Ackerson was in failing health and not expected to live.[23] With this new reminder of his own advanced age and mortality, Meeker and the Pathfinder prepared to head east.

CHAPTER 12

The Pathfinder Expedition, Part 3: 1917

East from Los Angeles

Marion Meeker's shift as a driver came to an end as the tour headed east, so Ezra searched out another driver and instead found two—brothers Edward and William Johnstone.[1] Meeker drew up a contract that the three men signed. The brothers agreed to pay for gasoline and repairs in addition to driving the Pathfinder to Washington, DC. They also were to help with camp chores. In return, they would receive 50 percent of the profits from the sale of Meeker's books and literature, and 50 percent of proceeds from his lectures. Where or how Meeker found the two brothers he did not say. The brothers made it as far as El Paso, Texas, where Meeker fired them.

The Pathfinder Company had been suggesting Meeker write an account of his trip around the country, and offered to pay publishing costs. Meeker only managed to write an account of the final leg of the journey. He had it typed but never published. Instead, it languished for decades in the Meeker Papers in a box that primarily contained his business correspondence.[2] It is nearly fifty pages in length. Excerpts follow, written by Meeker as the events occurred, in his uniquely digressive style, providing a first-hand glimpse of automobile travel in its infancy, on roads that were still little more than dirt trails. He titled it "Story of a Trip from Washington to Washington: A Swing Around the Circle, Eighteen Thousand Miles in an 80 Horsepower, Pathfinder Car." He began his story with two fibs:

"In another week, I shall, if I live that long be eighty-six years old—young I often say, for I feel it not having been a day sick in bed for sixty five years." He obviously forgot the month spent in a Seattle hospital in 1913.

"Why did you select a Pathfinder car I am often asked by parties who had never before heard the name? Perhaps the name attracted me. So many people had applied that name to my work…the very name had become familiar and associated memories of pioneer days kept alive." The real reason, of course, was that the Pathfinder Company was the only automobile manufacturer that was willing to build him a Schoonermobile.

Meeker planned to traverse the southern border of the United States and hoped to reach the nation's capital by February 1, 1917, a full month before the final adjournment of the 64th Congress. The Pathfinder departed Los Angeles at 3 p.m. December 26, 1916. The odometer read 10,945 miles. The car, gear, and passengers weighed 5790 pounds, 132 of which were made up by Ezra Meeker. This was down from the 165 pounds he weighed when he started west in 1852, and the 146 pounds he weighed upon arriving in Portland at the end of his first Oregon Trail trek.

The first day's run was 131 miles over a slippery road, mostly in the dark, to a San Diego hotel and to bed at midnight. The second day was 124 miles to El Centro where they made a night drive using a spotlight to supplement the headlights.

> Ed took the wheel and Bill, (as the brother Johnsons call each other) worked the light as a search light, first throwing the light on one side of the road, then the other, and again the center, to point out the rough places. That there was real danger we saw by upturned automobiles…and by a verbal report where an embankment gave way, a machine turned over and a man killed.

As they drove out of El Centro Meeker commented that it didn't look much like a desert as there were green fields along the road as far as he could see. But as they approached Yuma they "found the desert, the real thing, to remind us of scenes on the Plains in 1852; shifting sands, revive my memory of the dust storms prevailing at times." In one section of dunes a plank road was laid across the shifting sand. That night they camped by a desert oasis. Two days farther got the Pathfinder to Phoenix where they encountered "Fisks Auto Stage," which serviced customers within a 125-mile radius. Meeker wondered, if this could be done at a profit, why not a similar service in every city in the land or even a transcontinental service? With remarkable foresight

Meeker wrote, "In no very distant day.... If the national Government will build a Military Highway along the route of the Oregon Trail, or along the border-line fronting the Mexican border, these auto lines will speedily follow and become indeed and in fact a strong competition of the Railroads for certain traffic."

Sixty miles out from Tucson they came to the Gila River and found the stream had torn out the bridge approach. The state was charging one dollar to tow autos across the two hundred feet of running water. "I argued to myself what's the use of having an eighty horse power Pathfinder car and not use the power." A council with the two drivers was held and away they went. A three-foot hole splashed water to the top of the radiator but they made it through.

Bill smelled something burning on the way to Douglas, and the travelers discovered that a quilt Ed had been laying on while smoking was on fire. Meeker blurted out "Well why will any man smoke, it's a nasty filthy habit!" When the two smokers heatedly responded they "had a right to smoke," Meeker replied, "Well, you have as much right to commit suicide." Meeker seemed to have forgotten that he also smoked as a young man.

On January 4 they reach Deming, New Mexico, and the next day El Paso, where the U.S. Army had stationed fifty thousand troops. The report was "bad to awful roads ahead" for the next seven hundred miles. Here Ed and Bill insisted something was wrong with the automobile and that they needed to go to a garage, where they spent the remainder of the day. The two drivers overslept the next day and the next, and kept coming up with items that were in need of repair before they could start. After three days of this Meeker wrote, "Discharged Ed and Bill and now feel better," although he suspected them of sabotaging the Pathfinder.

An ad in the afternoon newspaper for a new driver brought in many applicants. Dan Lang was selected.[3] The bedding and covers had been washed, and books repacked when Lang discovered he had left his clothes at his mother's house. Another delay. Once on the road the two men found the road as bad as advertised. At times they actually had to drive in the Rio Grande riverbed. At Fort Stockton Meeker lectured to a small audience, "the most unappreciative one I ever stood before." The weather turned cold and wet and the duo traveled sixty miles without seeing a house and only two other automobiles. Eventually they got stuck, "the wheels would spin around but the car didn't move." Then they ran out of gasoline. They ate dinner and then began walking, and in two miles came to Ozona where they obtained "rescue." That night Meeker lectured to an appreciative full house. On January 22 the Pathfinder reached San Antonio, home of the Alamo, where Meeker had camped in 1912. Upon observing the people of San Antonio, Meeker wrote:

> It would seem that half the population of San Antonia was colored by observing the passing throng.... All shades of color from the mulatto of light complexion to the darker hues and then again to the point where one could not detect any colored strain of blood. A question came uppermost to mind, would there come a time when race prejudice would disappear; when the preponderance of the colored element would prevail to so large a proportion that equality of rights would be conceded, the "Jim Crow" [railroad] car abolished all over the south.

The drive to Houston and beyond proved rough. The spokes of one wheel were smashed, a tire needed repair, and the speedometer chain broke. Just east of the city they nearly became mired in a mud hole, then the rear wheel detached and spun off into a ditch, dropping the car body onto the roadbed. They hitched a ride back to Houston where they obtained the needed parts and equipment to repair the tire, took a streetcar to the end of the line, and walked four miles on a slippery muddy road back to the Pathfinder where they made the necessary repairs. They then drove at an eight-mile-an-hour pace to the next town. It was sixty-one miles to Beaumont and the general consensus was that they couldn't get through, but they did, over what Meeker felt was the worst road yet encountered. He was about to discover worse. "We all have heard of the last feather that broke the camel's back, but not of the last inch of mud that stalled the automobile. We came very near finding that inch today in several places where we were compelled to keep in the rut as a safeguard against skidding into the ditch."

To exit Texas the Pathfinder was loaded onto a barge that worked down a narrow waterway bordered on either side by low swampland, which hosted hordes of mosquitos that descended upon the travelers and broke the monotony of the trip.

After entering Louisiana they took to the road once again and at Vinton left the mud and struck a narrow brick road. They raised their speed from five miles an hour to thirty and passed through Lake Charles, when a wheel spun off the car. Wrote Meeker,

> Another walk of two miles after wire for temporary repair to reach Lake Charles again, the first back tracking of the trip. Three hours sufficed to make our own repairs and so with a night run of 12 miles to a camp close by the road side where, after midnight this writing is penned, and so I am off to bed, and soon will be asleep in the hope of better luck tomorrow, and thankful the mishap was not worse.

On January 31, they were told again "you can't get through" but this time the prophecy proved true.

> Dan made a mis-drive, I think, and ran two wheels off into a deep ditch. Not ten minutes before I had mentally congratulated myself that we hadn't been stalled on the whole trip from Seattle here—vain mental boasting as shown by the sequel. Half a dozen pulls down the ditch with one wheel in the ditch, which was filled with water, the other in the mud with the car at an angle almost to the rolling over point finally pulled out with no harm done. We hadn't driven half a mile 'til we plunged into a mud hole and stuck hard and fast and couldn't budge a peg. A neighbor farmer came with a block and tackle, and after breaking his rope three times, gave up the job. Then a resourceful young man came with a team of mules, and another block and tackle and worked diligently 'til nine o'clock, but failed to move the car an inch. This morning Feb 1st he sent seven miles for a cable and blocks and by doubling one set on to the other, brought the car to the surface at the noon hour after a day and an hour in the mud. We then pulled up to his house near the road, a mile distant, ate a splendid farmer chicken dinner, prepared by the mother, who said she knew we would get out of the mud about dinner time. When asked how much he charged, he said, "Nothing, I only wanted to help you along."

The trials were not over. The battery had run down as the generator "did nothing under a fifteen mile speed." Because of the bad roads they were compelled to run at low speed, and starting the car proved difficult. But they made it to Crowley where they sent the battery to a local shop to be charged. In the morning they were told the battery was dead and could only be "resurrected" by connecting it to the local powerhouse. They were two hundred miles from New Orleans. Then the weather turned cold, freezing the mud. In the afternoon of February 3 they started the twenty-eight-mile drive to Lafayette with a working battery and the generator adjusted to ten miles an hour. Again they became stuck, and with the help of a neighborly farmer and "two little mules that pulled like Caesar," made it to Lafayette not long after dark. "I begin to wonder how many more such experiences we are likely to have," wrote Meeker. "I usually pass unpleasant experiences with 'it might have been worse,' but that mud hole at camp 'Stuck in the Mud' stopped me from saying anything for in fact I cannot imagine how it could be worse.... If someone had come along and said it might be worse I almost believe I would have mentally called him a liar."

Unfortunately it was about to get worse. A stretch of "good road" tempted them to remove the tire chains they had been using to get through the mud. And late in the afternoon they plunged into a hole and stuck fast with no shovel—it had disappeared back at Crowley, either lost or stolen. They jacked the wheel up, put a board under it, and went to bed.

> Soon after we had gone to bed the north wind raised almost if not quite to the velocity of a gale, and sent shivering gusts through openings in the cover. We had no fear of overturning with the iron braced body absolutely fastened to the heavy chassis under us the whole of an aggregate weight of 4,400 lbs but strive as best I could it became impossible to keep comfortably warm. The temperature fell to below freezing, and ice formed by morning over a half inch thick.
>
> So when morning came, and breakfast served in the "dining car" we prepared for the start, when lo and behold the wheels wouldn't go round. Crank as much as we could, (for a twelve cylinder car is very hard to crank), the result was the same; the machinery wouldn't start. We had filled the radiator with hot water, but to no purpose. The transmitter from the generator (or visa versa, for I don't know which) had been set to indicate transmission only above fifteen miles an hour. Over the bad roads the speed was less and the battery run down. At Crowley a Mr. Walker, recommended to me a man of long experience, an expert in fact, said he could adjust it to generate at a ten mile speed. Dan said it would 'burn out" but Mr. W. said there would be no danger, and so I had him adjust it accordingly and seemingly it worked all right, but when Dan couldn't find any other point to prevent me from starting, he concluded his prophecy had proved true, and that the trouble lay in this particular place, and that it could not be replaced this side of New Orleans, ninety miles away—by rail. And so I ate an early dinner, walked two miles to Gibson, took the

train and arrived in New Orleans too late to do any business, except to read and answer a large accumulated mail received.

Meeker spent the day in New Orleans looking for the faulty generator part and a battery for their "electric torch." It was dark before he got back to the Pathfinder. The following day brought rain, and the car "skidded so bad it became a serious question whether we could safely drive at all." They stopped during one downpour at "a desolate cabin occupied by an eccentric man who was boiling something made up of a number of ingredients that gave it a repulsive color of the mud outside. He called it samp." They were reminded that they were but ten feet above sea level by the mud marks on the cabin wall that were two feet above the floor. It took them all day to go just seven miles.

They camped the next day in a garage where they found eight of their spark plugs broken. Meeker telegraphed to Indianapolis for the plugs, settled in his own mind to wait for the mud to dry up. The following day he took the train thirteen miles out, intending to walk back and examine the road that far. Immediately he found deep mud clinging to his shoes, almost pulling them off his feet.

> If the road was impassable for automobiles it is doubly so for pedestrians, and so after three miles of the torment, I took to the railroad, became a veritable tramp counting the ties, and after a ten mile walk, at dark reached the car, tired, hungry and with sore toes. After all the trip was not without results. The question was settled, I could not proceed 'till the mud dried up, at least in part. The sun was shining, and the north wind blowing, both favorable factors to dry the road.

At Donner, Louisiana, on February 10, 1917, Meeker wrote, "If possessed of the least taint of superstition, I would conclude some 'Hoodoo' spirit haunted the trip. As we drove out of the Garage yesterday the wheel…came off without any apparent cause, after having stood the test of several days driving over the worst of roads—in five rods of driving the gasoline was all gone."

The rain continued and Meeker surrendered. He made arrangements to ship the Pathfinder by rail to New Orleans. It took two hours to load the auto into the largest boxcar in the system and they had to let the air out of the tires to squeeze it inside. Upon arriving they found the influx of visitors for the approaching Mardi Gras celebration had filled all the hotels. Hours of searching turned up nothing. The Strangers Aid

Society referred him to the Southern Pacific waiting room, then to a hotel where he was assigned to a cot in a room with twenty-four others.

> I had slept with the indigents, and in a building where several hundred had slept, and paid ten cents for the privilege, because they could not afford to pay more. I was in, or of the submerged class—quiet, sober people, some with white locks, others of middle age. There was no effort to make the place cheerful; a dark room; could not find the number of the cot without striking a match, nevertheless I slept soundly, and up early next morning refreshed, but not feeling sure whether disease had been lurking in the room or not.

On February 15 the Pathfinder was shipped to Indianapolis and the arrangement between Meeker and the Pathfinder Company was terminated. Finances were so low at this point that Meeker had just enough funds to make it home. But to do so without making an attempt to induce Congress to build the military highway was unthinkable. The 64th Congress had adjourned long before he reached New Orleans, but a call had been made for an extra session of the 65th Congress, a sure forerunner of war. He wrote,

> If ever a Military Highway was needed anywhere, and at any time this border line road fronting the Mexican border was needed now; and so on the 23rd of March, took passage by rail and the 25th arrived in Washington City. The result of the effort to secure Congressional action will be recited later in this narrative, suffice it to say the approaching declaration of war precluded interest in any other question other than preparedness with the result of transferring this question to the Military arm where it rests at this writing.

On April 2, 1917, President Woodrow Wilson asked the United States Congress for a declaration of war against Germany. The vote in the Senate was 82–6; in the House of Representatives it was 373–50.

Shortly thereafter Meeker wrote Secretary of War Newton S. Baker urging him to support the immediate building of a military road along the southern border of the United States and also put in a plug for Pioneer Way.[4] He also wrote Senator Miles Poindexter, "You will doubtless remember that patriotic appeal of Lloyd George 'Too Late,' while addressing the House of Commons where awful sacrifice had followed from not being prepared in time. For Heaven's sake let us not follow in the footsteps of England and not make our preparation until it is too late to avert disaster or great sacrifice."⁵

After a jaunt to nearby Mt. Vernon Meeker took a fall on a city street and was briefly hospitalized. He carried a day calendar in his vest pocket that stated on the inside cover, "in case of accident contact Carrie Osborne 1120 38th Ave. N. Seattle and telegraph W.P. Bonney 401 N. Cliff, Ave Tacoma, Wash."⁶ Word of the "accident" quickly reached Puget Sound. Bonney immediately wrote to Senator Lewis asking what hospital Meeker was in.⁷ As this news spread, a letter from Meeker to his daughter, outlining his plan to hitchhike home, reached Seattle. Carrie was away for two weeks, so her husband Eben Osborne opened the letter and immediately sent Ezra "a draft for $100—for railway fare & sleeper to bring you home.... We want you home & I don't intend that you shall suffer if I can help it."⁸

Meeker caught the train in Washington, DC, on April 25, 1917, and was back home in Seattle by the end of the month. Another grand adventure was over, once again he was broke, and the focus of the country was now on war. For a time Meeker was forced to put his plans for "Pioneer Way" on the back burner, but they would not stay there for long. In the meantime, he too turned his attention to the war effort.

The Pathfinder Company died one year later according to the September 1918 issue of *Automotive Industries, The Automobile: A Journal of Automotive Engineering and Construction.*

"Indianapolis, Sept. 15—The plant of the Pathfinder Automobile Co., which recently passed through dissolution proceedings, has been leased by the S. M. Dixie Shoe Polish Co. of Brooklyn, NY and will be used for the manufacture of shoe polish."

A sad end for a "mighty car."

CHAPTER 13

World War I

Once home, Meeker quickly changed course. He wrote William Bonney that, "duty calls every American citizen to encourage food production, utmost possible.... I am thoroughly convinced the 'man with the hoe' is as important in this crisis as the man with the musket and believe I can do something and that I ought in the line of food production and preventing waste, hence this change in my program."[1] By June Meeker was shopping for small portable food dryers and seeking information from a Canadian company on the latest methods of drying food.[2] He wrote to a Wisconsin company inquiring about shipping containers.[3] While Washington State was still a large producer of hops, the crop on which Meeker had made and lost a fortune,[4] the transition in Puget Sound to berries, fruits, and vegetables was well under way. Meeker knew that Washington could produce huge quantities of fruits and vegetables for the war effort. All that was needed to get a mutually beneficial boom under way was to secure the right government contracts. British Columbia was sending thousands of tons of food products to the English government and Meeker saw a similar demand likely to come from the United States government. He sent letters to members of Congress and to Herbert Hoover,[5] head of the U.S. Food Administration, pointing out the Northwest's ability to aid the war effort.[6]

True to character, Meeker was indefatigable in pursuing his new project. He requested that Washington State Senator Wesley L. Jones connect him with army and navy purchasing agencies. He also approached the Seattle newspapers with a proposal to write a weekly column about growing "War Gardens," and asked the Seattle City Council to set up a market where the surplus of the multiple gardens in and around the city could be brought for shipment overseas. He also proposed starting a business to collect, preserve, and ship foodstuffs in partnership with The Dalles, Oregon, Wittenberg-King Company:

149

I will now proceed to incorporate and set out the object solely for the benefit of the Allies, the sufferers of France and Belgium and our "Boys" in the trenches.... I evaporated several hundred tons of the fresh vegetables 1898–1900 for the Klondyke market but we cooked most of this before drying but some we dried uncooked.... Our people do not yet realize the serious crisis that confronts us. I believe there is a long war ahead of us. I am willing to devote my whole energies to aid in the struggle.[7]

Carrie and Eben Osborne had a summer home on the east side of Lake Washington at Yarrow Point, and Meeker started his own garden there in which he planted American Wonder potatoes, corn, and strawberries.[8] By August he also had three fourths of an acre of beans under cultivation.[9] He asked Charles Hood of Puyallup if he could round up four acres in that city for a garden, "For our Washington Boys in the trenches."[10] To Herbert Hoover he presented the idea of dehydrating vegetables on a national scale. He wrote that the state would produce a large surplus of perishable fruit and vegetables that should be dried, as canning was too expensive and cans or glass unobtainable. He went on to tell Hoover that he was sending detailed plans via Senator Jones. He wrote,

> My friend, I am an old man—nearly 87, but yet hale and hearty and able to do my "bit" and my services are at your command if perchance you need me. I have this year cultivated a large garden with my own hands, the resulting crop to go for the benefit of the cause, and am undertaking preparation for a larger one next year.[11]

On August 9 Meeker traveled sixty miles north to Arlington where he gave an address, complete with his motion pictures, to the Pioneers of Northern Washington.[12] The Arlington newspaper published the long address in which Meeker told stories about the early pioneers and ended with a call to plant gardens for the war effort.[13]

Several of Ezra's grandchildren were directly involved in the war effort. Bertha Templeton had accompanied Meeker for several months in 1906 while he was on his Old Oregon Trail Monument Expedition, and she worked for him in 1909 at his restaurant at the Alaska-Yukon-Pacific Exposition, and in San Francisco at the 1915 Panama-Pacific Exposition. When she took a job with the war department in Washington, DC, in late 1917, Ezra gave her some contact names and hints on where to find a room.¹⁴ Meeker was also quite close to his granddaughter Olive Osborne, Carrie's daughter. Olive had helped care for Eliza Jane Meeker during her final illness. She transcribed much of her grandfather's literary efforts, graduated first in her class from Whitman College in Walla Walla, Washington, and became a RN at Cooper Hospital in Camden, New Jersey. When the United States entered the war she joined the Army Nurse Corps and in April 1918 was ordered to France.¹⁵ Ezra was immensely proud of her and showered her with praise.¹⁶ While in France, Olive met Captain Meryl Jones, whom she married on March 10, 1920.

Meeker spent 1918 working primarily on two projects. The first was the war garden he planted at the Osborne summer home. In his first year of gardening, he cleared seventy-five dollars which he then turned over to the Red Cross. In 1918 he cleared $257.80, the product of an acre and one-third. A *Seattle Post-Intelligencer* reporter gave the details of Meeker's "Victory Garden," writing,

> As an incentive, if one were necessary, Mr. Meeker had a sign nailed up at one end of the garden, reading: "This Garden is Dedicated to the Defenders of Liberty." If the spirit lagged during the heat of the day after hours of hoeing, a statement that anyone knowing Mr. Meeker will challenge, all he had to do was take a look at the sign and attack the enemy weeds harder than ever. "It was a pleasure and a duty," insists Mr. Meeker, "and I do not think anyone should take credit for doing what he considers a duty."¹⁷

Meeker's second project was his proposed food drying enterprise. He had perfected this method of producing potato flour during his Klondike years and he offered the recipe to all who asked. He lobbied Herbert Hoover and other government officials to convince them that dried potato flour could substitute for wheat flour, which had become scarce in Europe due to the ravages of war. For two months Ezra worked to apprise the authorities in Washington, DC, of the importance of conserving Washington's 1917 potato crop, now harvested and sitting in warehouses. In March 1918 Meeker met with Charles Hebbard, the federal food administrator for Washington State, at a Chamber of Commerce gathering. Ezra assured him that financing could be handled locally provided the government assured the manufacturers of a market. Three times Herbert Hoover replied to Meeker, misunderstanding the request and stating that he had no money to build plants. Meeker wasn't asking him to build plants. He was confident that the various hop kilns that still dotted Puget Sound could serve adequately for drying the potato flour. He was simply asking for funds to outfit them for processing flour and assurance that once dried, the government would purchase it.[18] Ezra turned to Senator Jones for help,

> If this war is lost, it will be on the food question, and yet we are confronted with the appalling fact we are likely to lose the product that would produce thousands of barrels of flour and discouraging farmers from planting for the 1918 crop. It is indeed a very serious situation and if we lose a part of this crop [we] will feel cast down and discouraged. Our enemy you know would not allow such a waste; is it possible that something cannot yet be done?[19]

In early November 1918 Meeker finally received word via telegram from Senator Jones that the food administration had arranged to finance a plant. Unfortunately the news came too late to be useful. World War I ended on November 11, 1918, and with it came the demise of Meeker's dehydrated potato plan, although he continued planting his garden throughout the following year.[20]

After the war, the Treaty of Versailles and formation of the League of Nations inspired much discussion within the Meeker family.[21] Meeker's daughter, Ella Templeton, who was in Washington, DC, in July 1919, wrote her father about attending the Senate treaty debates. She did not offer an opinion, but the topic much agitated Meeker. He engaged in a spirited correspondence with his nephew, Ella's son, the Reverend Harry Templeton, arguing that by joining the League of Nations the United States would be surrendering its sovereignty and independence to a confederation. Meeker viewed this as dangerous. Marion joined in the debate from California, siding with his father. However, son-in-law Roderick McDonald saw value in the League, perhaps because his own son was in Vladivostok with the Canadian military as part of the allied forces involved in the 1918–20 Russian Civil War.

As the decade neared its end, Meeker returned to his grand quest to save the Oregon Trail. He renewed his effort with a focus on his home state of Washington and the Naches Pass Wagon Road.

CHAPTER 14

The Fight for Naches Pass

With the war over and enthusiasm for Meeker's Pioneer Way waning, Ezra reached back in time and centered his focus on another aspect of the Oregon Trail. As always, he poured everything he had into the project.

In the summer and fall of 1853, under the guidance of Edward Jay Allen, the citizens of Washington Territory built a rough wagon road over Naches Pass, a cross-Cascades route north of Mount Rainier. Its purpose was to siphon off a portion of the immigration bound for Oregon's Willamette Valley and divert it to Puget Sound.[1] In October of that year the Longmire-Biles wagon train became the first to attempt the new road. Their journey over Naches Pass has since become a northwest legend. The climax of the story occurred when wagons had to be lowered by ropes down a cliff. At least one wagon did not survive the descent. Beefing up the tale, pun intended, is an apocryphal story of the travelers killing three oxen to make rope from their hides, a story repeated in Meeker's book *Ox-Team Days on the Oregon Trail*.

> James Biles, one of the leaders commanded, "Kill a steer." They killed a steer, cut his hide into strips, and spliced the strips to the rope. It was found to be still too short to reach the bottom. The order went out: "Kill two more steers!" And two more steers were killed, their hides cut into strips, and the strips spliced to the rope, which then reached the bottom of the hill.[2]

For years Meeker had been telling the tale of "Kill another ox!" to audiences around the nation. At other times he added to the tale by recounting his own journey over the pass in 1854. With the dawning of the automobile age upon the country, Meeker did his best to turn Naches Pass into a *cause célèbre*.

A Brewing Conflict

The first automobile went over Snoqualmie Pass in the Cascade Mountains in 1905, the year the Washington State Department of Highways was created. The rough track was upgraded in 1909 for the Guggenheim transcontinental automobile race that was one of the highlights of the Alaska-Yukon-Pacific Exposition. In 1915 a new two-lane road, the Sunset Highway, was built over the pass, connecting eastern and western Washington.

Two years earlier, in 1913, the Washington State Legislature debated the idea of constructing a second highway over the Cascade Mountains somewhere south of the existing highway through Snoqualmie Pass. They looked at two possibilities. The first was to follow the ancient trail over Naches Pass that became the famed wagon road of 1853–54, and in later decades a stock trail from the Yakima Valley to Puget Sound. The second possibility was to go over today's Chinook Pass, then called McClellan Pass. Lined up behind this choice were the Great Northern Railroad, the Rainier National Park Concessionaire (RNPC), and the National Park Service. They envisioned a major tourist resort at what is today's Sunrise. The legislature chose the McClellan Pass route as it gave the people of eastern Washington and Seattle a direct route to the northeast entrance of Mount Rainier National Park and, at this point, to the simply dreamed-of resort. To cement this decision in 1913 the state constructed Normile Grade, a six-mile section of highway east of the pass at a cost of $80,000. This piece of highway connected to nothing. It was simply a guarantee for the future. To further lock in the choice, Yakima County improved a mining road that went from Bumping Lake to the Cascade Crest at Tipsoo Lake. When the state resumed road construction on the west side of the mountains in late 1920 Meeker decided to fight for Naches Pass. He wanted this route to become part of his Pioneer Way and for several years he turned his energy and attention to that end.

Retracing the Naches Pass Trail

While Meeker was away from the state on his second Oregon Trail Monument Expedition, the work of remembering local history fell to others. Early in 1910 William H. Gilstrap and George Himes, the secretaries of the Washington and Oregon State Historical Societies respectively, met to plan the placing of a monument to mark the last camp of the Longmire-Biles wagon train on Clover Creek in what is now Parkland, Washington. On July 14, accompanied by William Lane and Van Ogle, both members of that wagon train, Himes and Gilstrap located the spot and in due course a monument was erected. Gilstrap then arranged to have David Longmire, another member of the 1853 wagon train, take Himes and himself over the Naches trail from the last crossing of the Yakima River to the site of the last camp. Longmire lived on the east side of the Cascades in Wenas, a farming community on the old Naches Trail. He had driven stock over the trail for years and was quite familiar with the route.

Longmire met Himes and Gilstrap in North Yakima and drove them to his home, where Longmire's four sons joined the group. They took a wagon from Longmire's ranch as far as the junction of the Bumping and Naches Rivers. From there they traveled via horseback with three pack animals carrying their supplies. They left the Bumping River camp early on August 21. By noon the next day they had reached the famous Naches cliff down which the 1853 wagons had been lowered with ropes. It took four and one half days to cover the old trail from the Longmire home to Buckley, a distance of 126 miles.[3] Himes kept a diary of the trip and Gilstrap took some twenty pictures.[4]

For the next eight years there was little talk about road building in Naches Pass. Then in early 1919 Meeker suggested a repeat trip to Naches Pass to Longmire and Himes, along with Seattle pioneer and historian Clarence Bagley, and William Bonney, secretary of the Washington State Historical Society. Meeker, now eighty-eight years old, had last been over the pass in September of 1854 when he guided his father's wagon train from the Columbia River over the mountains into Puget Sound. For Himes (seventy-five) and Longmire (seventy-four) this would be their second trip over the route in the decade. For Bagley (seventy-six) and Bonney (sixty-three) it was new territory. It was to be a camping trip.

After much planning, the men agreed to meet in Steilacoom, Washington, on July 11. The plan was to verify the route of the old wagon road as far as possible, and to put up temporary markers at different points along the way commemorating the journey of the pioneers of 1853 and 1854.

Himes told the would-be explorers:

> To follow the exact trail of 1853 is an impossibility—that is, relatively. The general course of the present trail made by the forest rangers is the same. Even the road built under Lieut. Arnold in 1854, supervised by Edward J. Allen, who was largely the one who "engineered" the trail of 1853, diverged frequently from the original road. David Longmire is the best authority upon that point, as he and his father went over it two to four times a year from 1859 to 1884. I will have my diary of the trip in August 1910 with me as well as the photos taken by Gilstrap.[5]

Himes almost missed the journey as he was struck down with what he described as a "preposterous attack of lumbago" just days before his scheduled departure from Portland. He was afraid that Meeker would "crow like a ha-China fesant" if he failed to make the trip and so summoned his body to get well quickly.[6] There would be no crowing by Meeker.

The party met as planned in Steilacoom for that city's annual picnic where Meeker gave a speech. The next morning they left Tacoma by automobile and made their way to Enumclaw, stopping first at Clover Creek to view the monument placed by Himes and Gilstrap in 1910. From Enumclaw they made their way to just east of Greenwater and camped for the night at a grassy area called Bear Prairie. According to Meeker's day calendar, the next two days were spent tramping eight miles up the Greenwater River, which they crossed five times on foot logs. They made it to within three miles of the summit before Bagley and Himes surrendered to their years and insisted on returning. The party retreated and camped again at Bear Prairie.

On July 15 the old timers were back in Enumclaw. They stopped for a time at Porter's Ford on the White River and camped at Montgomery's in what is now Spanaway. The next day they were at Yelm being hosted for dinner by Longmire family members. That evening Meeker returned to Seattle while the other old-timers camped on the site of the 1853 Himes family claim in today's Hawks Prairie in Lacey. They then drove over Snoqualmie Pass to Yakima to examine the eastern end of the Naches trail.[7]

The next summer Meeker returned to Naches Pass. He was traveling in the Yakima Valley when he heard that William Anderson of Naches was going to lead a remuda of thirty horses over the Naches

Pass trail. The Seattle Mountaineers had rented them for their annual summer outing in the Olympic Mountains and Meeker jumped at a chance to go along. His trip the previous summer failed to reach the summit and here was a chance to complete that unfinished business. It also offered adventure, something Meeker rarely let pass.

On July 15 George Chambers of Wenas drove Meeker to the William Anderson ranch on the American River after making a side trip to Boulder Cave. Howard Green, also of Naches, vice president of the Cascadians, a Yakima, Washington, outdoor recreation club, accompanied Meeker. The horse packers had dinner at the American River and started up the trail. That night they camped at Crowe Creek. The next day they climbed to the summit meadows at 4,900 feet and camped. Most of the distance was made on horseback until the trail became too steep and narrow near the summit. At that point "the aged pioneer essayed the most difficult part of the journey afoot."[8] The *Yakima Herald* wrote, "It would naturally be thought that slow traveling would be the order of the day with the handicap of the aged man, but such was not the case."[9] According to Anderson, Ezra usually led the procession with his white hair flowing out behind him in the wind, only dropping back when he wanted to inspect old markings and grades. They actually found some old moss-covered logs that the pioneers had chopped sixty years earlier.

Lunch at noon on the third day was spent at the upper crossing of the Greenwater River at a ranger station manned by Joseph W. Albro. Here a fatigued Ezra rested on the ranger's bed. After a short nap he was up, refreshed and ready to continue. At 6 p.m. the party arrived at the Crystal Ranger Station near the mouth of the Greenwater River. The only casualty of the trip was a lost watch that Ezra had carried for thirty years. (The watch was eventually found and resides today in the collection of the Puyallup Historical Society at the Meeker Mansion.)[10] An automobile met them the next morning at Greenwater and returned Ezra to his Seattle home. The horse trip covered some thirty-five miles of pristine trail. Meeker summed up the trip for the *Seattle Post-Intelligencer*:

> The old trail seemed narrower, the mountains higher, the rivers wider and deeper and the switchback turns sharper than in 1854. It amazes me to realize that we ever got ox wagons and teams over the pass trail in those days. It may be that I am getting old that made it

seem that way. But I can still show a lot of these young bloods how to cross the mountains. They told me I couldn't do it, but I did. The only bad effect it had on me was to work up an appetite.[11]

Meeker ended the interview by saying he would organize a party soon to return to the pass to search for the old tree used in lowering the teams and wagons down the more than a half century before. Pieces of the tree, if found, would be used for exhibits in the Ferry Museum. He told the Tacoma newspaper that a man in Greenwater knew the location of the tree and was willing to guide a party to it.

COMPETITION FOR THE SOUTHERN CASCADES ROAD

A third contender for a southern road over the Cascades appeared in 1920. Senator Phillip H. Carlyon, of Olympia, envisioned a route much father south, over today's White Pass. He introduced a bill in the legislature that authorized the expenditure of thirty million dollars to pave one half of Washington's 1,516 miles of roadways. Included in the bill was funding for the construction of a road over the newly named Carlyon Pass.[12]

Meeker fought this bill vigorously and began a statewide tour campaigning for Naches Pass as the proper route for the southern highway. He argued in the press that $35,000 per mile as called for in the Carlyon Bill was grossly expensive, that the lifetime of concrete paving was too short to justify the cost, and that the issuance of bonds in this amount was a debt the state should not incur. In a letter to the editor of the *Centralia Chronicle* Meeker wrote of the foolishness of spending $130,000 on the McClellan Pass road up White River from Greenwater arguing that this road to Yakima headed straight to an impassable barrier of snow. He pointed out that in 1913 the state built a six-mile section of road just over the crest of McClellan Pass at a cost of $80,000—a highway that "never has been traveled, began nowhere and ended in a snowbank, ten feet of snow on the 20th of June, 1920."[13] The *Centralia Chronicle* told more.

> He is fresh from a "jaunt" over the famous Naches pass in the Cascades where he found 38 miles of the 135 mile stretch between Yakima and Auburn in the same primitive condition it was 66 years ago. And Ezra is just plain "mad" at the governor, the legislature and all the rest of the politicians because the state is spending millions of dollars on other passes that he claims are naturally inferior for cross-

state roads. And he is telling automobile clubs, commercial clubs, and other civic bodies so in no uncertain words. His present tour of the west side will take up all his time between now and August 25, the date set for the meeting of the Good Roads association in Everett. He will leave Elma tomorrow for the west end of the county, planning on arriving at Aberdeen Friday of this week.[14]

In this Meeker was partially successful. By November the Carlyon Bill was dead, but the building continued on the highway over "inferior" McClellan Pass.

Promoting the Naches Route

Ezra Meeker began 1921 by writing to legislators about the Naches Pass road from his lodgings at the Fairfield Hotel on 6th and Meridian in Seattle. He pointed out that on the west side of the Cascades roads went from Seattle and Tacoma to the mouth of the Greenwater River and on the east side of the Cascades a road went from Yakima to the Bumping River, and that the state had expended $500,000 to date in constructing these roads. He encouraged the legislature to appropriate funds to survey the gap between the two ends of the current road. He argued that the Naches Pass route from Yakima to Tacoma was nearly seventy miles shorter when compared with the Snoqualmie Pass route and that it was also thirty miles less to Seattle. Meeker wrote that the Naches route could be kept open all winter and that only eight miles of it would be what he termed an "expensive mountain road." Meeker enclosed a map outlining the route with his letter.[15]

At the end of the month he wrote a letter to the editor of the *Seattle Post-Intelligencer* advocating a $10,000 appropriation to survey the "rough" eight miles near the Naches summit to determine if building such a road was cost effective. He added to his claims of it being the shortest route by stating it had the easiest grade of any path over the Cascades and that its average elevation was lower than that of Snoqualmie Pass. He suggested building a tunnel under the eight miles near the summit, similar to the one being contemplated near Denver in the Rocky Mountains. Meeker touted the economic benefits of connecting the commerce of the eastern half of the state with that of the western half. He pointed out that tourists coming to Puget Sound by automobile from the Yellowstone area would avoid spending their dollars in the Columbia River route and in Portland if the direct route over the Naches Pass was open.[16] All this prompted a trip to Olympia where he made his case in person to the state legislature.[17]

In a second letter he offered the readers of the newspaper a history lesson telling the story of building the 1853 wagon road and of the first wagon train to come over that road. Meeker wrote that some were arguing that now that the Snoqualmie Pass road was open nothing more needed to be done. "We heard that same contention when the first line of railroad was built," he wrote. "In fact, we long contended that a railroad could not be operated over the mountains in the winter; just as we now hear that a state highway cannot be kept open for winter travel. Now we have three railroad lines instead of one, all regularly operated and a call for another to be built in the no distant future." He went on to say that Senator Jones had a bill pending in the United States Senate appropriating money to "begin work to open the Oregon Trail as a military road," and that Jones' bill was likely to soon become law. "Shall we supinely stand by and see all the great tourist traffic so soon to throng that ancient highway go down through the Columbia gap and make no effort to prepare the shorter way to Puget sound and show the travelers the wonders of our Inland Empire?"[18] That same day he received a telegram from Senator Jones agreeing to change the routing of the military road in his Oregon Trail bill to Naches Pass rather than down the Columbia River to Portland.

A month later Meeker wrote to David Longmire that the legislature had adjourned without appropriating funds for the Naches Pass survey. He laid the blame on the Seattle members of the Washington State Senate. Meeker told Longmire that he would lead his own surveying party to examine the route, issue quarterly bulletins educating the public on all points of the proposed work, and carry the fight to the next session of the legislature. He concluded by stating, "It's a big job, but it's of great importance to the state and particularly of Yakima and if it succeeds would enhance the value of your own and son's holdings thousands of dollars. Now do you want to fight or do you say give it up."[19] On May 15 Ezra wrote William Bonney suggesting another personal trip to the Naches Pass area to do some surveying. Obviously, Meeker did not intend to give up

In August 1921 Meeker traveled to Enumclaw to attend the Farmer's Picnic and to meet and talk with some of the old pioneers. But he also took time to present his case for building a state highway over Naches

Pass to the Enumclaw Commercial Club. Ezra brought David Longmire and an engineer from Puyallup along to support his argument. His plan was to stir up interest in commercial clubs in Auburn, Sumner, Puyallup, and Yakima and encourage them to pressure their legislators.[20]

In November he was still working at it, writing a long article that again recounted the history of the 1853 wagon road, touted the economic advantages of the route, and claimed the U.S. Senate was close to appropriating funding for a military road along the Oregon Trail. He contended that if the current generation had the will of the pioneers of 1853, a paved road over Naches Pass would soon exist, and lamented the transfer of tourism and commerce to Oregon that would be lost if no such road was built. He concluded by stating, "The answer rests with you of this generation."[21]

Meeker continued his push for a cross-state highway over Naches Pass into 1922. In March he sent a letter to chambers of commerce and commercial clubs around the state pointing out once more the advantages of the Naches route and arguing that the proposed Carlyon route over today's White Pass would direct tourist trade south toward Portland rather than toward Seattle.[22] He also began soliciting information on the snow conditions at McClellan Pass, hoping to prove the late season snow melt made the route impractical.[23] He optimistically told Bonney he intended personally to lobby the legislature to begin construction of the final forty miles over Naches Pass. He also asked Bonney to feel out Governor Hart's position on the Naches Pass road, as the governor had previously been opposed to building a highway there.[24]

Meeker spent the second half of 1922 and most of 1923 out of state working on his feature-length Oregon Trail motion picture. With an eye to the Naches Pass campaign, one of the scenes Meeker intended to replicate in his movie was that of lowering the wagons down the Naches Cliff. He suggested there would be no film work done north of the Columbia River if that scene were omitted. Meeker intended to clear the original route of underbrush and film two wagons being lowered down by ropes. One of the wagons would be let go to smash as it tumbled down the grade. He also intended to kill three steers and cut their hides into strips, perpetuating the "kill an ox" myth. "I think we might develop a barbecue of the beeves, on the occasion, and have more people there on the summit than ever assembled at that point at one time, or probably ever will again."[25]

Meeker estimated the cost of filming the Naches Pass story would be at least $1,000, but he was unwilling to spend a cent in Washington if the Naches Pass road remained closed. He felt that if Governor Hart continued to oppose the opening of that route, it was useless to lobby James Allen, the head of the highway department, and that a legislative endorsement was needed instead. Bonney replied that although he had been invited to testify before a legislative appropriations committee, his talks with several Tacoma businessmen suggested taking the campaign to the legislature was also pointless. Bonney also informed Meeker that the chambers of commerce across the state opposed any appropriation for new roadwork.

Taking the Fight to the Legislature

In June 1924 Meeker was back in Seattle, charging that state officials were practicing deception by pretending to improve the Naches Pass route when in reality they were building the highway over Chinook Pass, sixteen miles distant.[26] Highway Commissioner James Allen clarified what was already pretty much a "done deal." The Legislature had considered the two routes in 1913 and decided upon the McClellan Pass (Chinook) route. The thousands of dollars that had since been expended would be wasted if the legislature now decided to select the Naches Pass route for "sentimental reasons." He further stated that the highway was built to within nine miles of the summit on the western side and twenty-seven miles on the eastern side, and announced that he was issuing a contract on June 25 for the construction of four more miles on the eastern side of the Pass.[27]

The deception, as Meeker saw it, was that the legislature had recently changed the name of the project to the Naches Pass Highway, misleading supporters into believing that the actual Naches Pass route was being used. It wasn't. Meeker saw this as akin to bait and switch.[28] Many of the old settlers and pioneers took exception to the name McClellan being used for either the pass or highway, and supported following the original route of the trail.

Meeker ultimately decided that he needed to be inside the legislative decision making body if he was to succeed. Accordingly, he announced his candidacy for Seattle's 47th District in the House of Representatives.[29] The central focus of his campaign was Naches Pass.[30] In an unusual move he ran a statewide campaign for a local legislative seat.[31] In mid-July he was in Spokane campaigning both against the state plan and for a position in the state legislature. On August 11 he flew from Seattle to Rochester to attend the meeting of the Southwest Washington Pioneer Society where he lobbied both for President Calvin Coolidge and Naches Pass. This was followed with trips to Longview and Vancouver. Apparently Meeker reached many of his 1924 campaign destinations by airplane.[32]

The money behind the Chinook Pass route proved a serious obstacle to Meeker's goal. The Great Northern Railroad and RNPC had pushed hard for a tourist hotel on the north side of the national park and viewed a state highway as central to that project. This push, likely coupled with money in various forms, brought local politicians onboard. The chambers of commerce at the portal ends of the highway were enthusiastic Chinook Pass supporters. They viewed a park hotel with its abundance of tourists as more beneficial to their communities then potential Naches route traffic. In addition, Yakima County had expended thousands of dollars pushing a highway toward Chinook Pass and Tipsoo Lake. John Archibald McKinnon, Enumclaw's state representative, teamed with the Yakima delegation and both were working to get the state to pledge to open the road over Chinook and Cayuse Pass. Only eight miles remained on the west side to connect with the track the Yakima citizens had constructed on the east side. Representative McKinnon stated that it would require an expenditure of $500,000 to build these eight miles; of course, ultimately the twenty-mile stretch of mining road on the east slope would have to be reconstructed or entirely replaced to meet state standards. Meeker's opponents correctly argued that the Naches route would not tap the north entrance to Mount Rainier National Park. The *Seattle Star* succinctly noted, "Anyway, it promises to be a warm night when Meeker and McKinnon open up at the next legislature."[33]

Meeker countered by making a trip to and through Chinook Pass from the mouth of the Bumping River to within two and one half miles of where the White River Road led into Mount Rainier National Park accompanied by two Yakima citizens. He noted that from the Bumping River to the summit of Chinook Pass the "road" was a single track that dodged around big trees and was not graded. It had many steep sections and sharp curves up the mountainside, which was almost bare of timber and prone to avalanches. He viewed this section of the route as a death trap and dared James Allen or Governor Hart to dispute this statement. Meeker argued that this road would never be used for commercial purposes and that it was simply a dangerous tourist road. He stated that the Naches route was twenty miles shorter and 1,700 feet lower, if a tunnel was constructed, making it a practical, all-winter road free from avalanche danger, and it was a direct route between the centers of population of eastern and western Washington. He went on to say, "It will cost no more to build the road through the Naches Pass, including the tunnel, then it would cost to build over the Chinook with its twenty miles of additional expensive road and some of it exceedingly dangerous."[34] On September 6 the ninety-four-year-old pioneer was on a soapbox at the foot of Pike Street in Seattle making one last pitch to the voters. The *Seattle Star* printed a photograph of Meeker on that occasion and commented, "Street speaking is no snap for a young man, but that doesn't daunt aged Ezra Meeker."[35] On September 12, 1924, Meeker lost a closely contested primary election contest by six hundred votes. The fight was nearly over.

In December Meeker made another effort in Wenatchee at the Good Roads Association Convention. He came to the convention with the endorsement of the Seattle Chamber of Commerce, asking simply for a survey of the Naches Pass route, telling delegates "we cannot live on scenery alone," and asking them to "remember there are millions of tons of farm produce in the Yakima valley in the future to be moved to Puget Sound ports."[36] The *Seattle Post-Intelligencer* reported that Meeker had "made his big fight today for recognition of the Naches Pass as an East-West route across the state and—lost." Meeker's comment, "Well, the fight isn't over yet."[37] He gave the effort one last try when he testified before the Washington House of Representatives' Roads and Bridges Committee where he urged them to adopt his Naches Pass plan. The *Post-Intelligencer* reported the outcome. "The committee heard Meeker's argument, but took no action, which is tantamount to approving the present location of the highway."[38] The fight for Naches Pass was over.

It took several more years to complete the road, known today as State Route 410. It wasn't until 1931 that the east and west roads over McClellan Pass (Chinook Pass) met at Tipsoo Lake. However, Meeker's idea of a Naches tunnel remained on the legislative table through the early 1930s. Fifty thousand dollars was even appropriated for a feasibility study. In 1959 the tunnel concept was codified into Washington law. In the early 1960s Governor Albert Rosellini established a committee to study the feasibility of building a toll road over Naches Pass. The report was favorable. In 1970 the Naches Pass route officially became State Route 168. Two reasons were cited for giving the route state highway status. The first was that SR 168 would be an all-weather highway. The second was that SR 168 would allow commercial vehicles to bypass Mount Rainier National Park where they are prohibited. The Naches Pass highway exists today only in theory.

Meeker was right of course. Today both Chinook and Cayuse Pass remain closed all winter. The state found the cost of keeping them open in winter and the danger of avalanche prohibitive. And, as Meeker predicted, this route is strictly a tourist road. No commercial trucking is allowed on these roads as they pass through Mount Rainier National Park.

The White River Road that opened off Highway 410 to Yakima Park (today's Sunrise) in Mount Rainier National Park is a major attraction for tourists from both sides of the state, as Meeker's opponents predicted. On any summer day the parking lot is packed with the vehicles that bring tourists to gape at the view or to hike the alpine trails. Today, instead of a hotel, there is a day lodge with a small cafeteria and gift shop. The old campground has long since been closed, as were the two hundred cabins built before the collapse of the hotel project; slowly the land is returning to its natural state.

Naches Pass looks much like it did in 1853 and 1854, albeit with second growth trees and logging roads aplenty. It never became a cross-state highway. Instead it has become what the forest service calls one of the nation's premier jeep roads. Meeker would still recognize places like Government Meadow or Pyramid Peak from whose summit Edward Jay Allen, the builder of the 1853–54 wagon road, placed a large American flag to welcome the pioneers into Washington Territory.

CHAPTER 15

Collaboration

Howard Driggs

In September 1920 Ezra Meeker met Howard R. Driggs, who was visiting Washington State while gathering information for his New York University doctoral thesis on the teaching of English education.[1] This meeting changed the trajectory of Meeker's remaining years. Throughout his life Meeker associated with people who made things happen. The list was long. It included railroad tycoons such as Jay Cooke, governors and legislators too numerous to mention, five United States presidents, numerous newspaper editors across the country, and personages such as Susan B. Anthony. The addition of Howard Driggs to the list led directly to Caspar W. Hodgson, president of the World Book Company, and that in turn led to George Dupont Pratt, who financed Meeker's film-making ventures.

For years Meeker had been visiting local schools where he found the children eager to hear his stories of pioneer days. It was on one of these visits in Seattle that Ezra met Driggs. Wrote Meeker, "For some years he had been working to preserve the stories of the pioneers in truthful and attractive form for the children of our country. The immediate result of our meeting and of several conferences that followed was a plan to collaborate in preparing and publishing the story of my life."[2] The September 20 entry in Meeker's day calendar reported that the two men spent several days looking at local historic sites, and that while at Puyallup High School they received a standing ovation. A partnership was formed during these visits that would last the remainder of Meeker's life.

Apparently Driggs was still in Washington in November as George Himes wrote to Meeker on November 16,

> I will try to go to Vancouver today and see Prof. to whom you refer, whose name I cannot make out because you have at last began imi-

166

tating the handwriting of your old friend Horace Greeley. Your ordinary writing has no terrors for me, but when it come to proper names, they cannot be guessed at unless they belong to the plebian class such as Smith, Brown, Jones, Bagley, including Meeker and Himes. I guess the Prof's name is Driggs.

Four days later, Himes reported that the men had met, and had gone together to Portland and "spent three hours in Historical Society rooms, selected a number of subjects that he thought would be useful in illustrating your book.... Will hear more of him later."[3]

After spending a portion of the fall of 1920 in the Northwest Driggs returned to New York and immediately went to work on the book about Meeker's life that the two men had contemplated in September. By late January 1921 Driggs wrote Meeker, "I am practically through with the new book now. It has taken me every moment of the time I could spare since I reached New York; but the effort will be well spent, I feel sure." Driggs aimed his writing at a seventh to eighth grade level, thinking that such would have appeal to the general working class, but kept faithfully to Meeker's style and spirit. He eliminated only those parts of Meeker's story that would appeal only to local interests or that he deemed extraneous. Driggs had difficulties in reducing Meeker's complicated sentence structure to something more readable and in reorganizing the material to make the story flow. He felt he had accomplished this and was ready to engage an illustrator. But he had a few questions he wanted Ezra to clarify involving vocabulary—for example, what was a puncheon floor?[4]

Despite doing the majority of the work on this book, Driggs saw to it that Meeker received a contract with the World Book Company and a 5 percent royalty.[5] Always the businessman, Meeker had his son-in-law, attorney Eben Osborne, review the contract.[6]

Meeker supplied Driggs with the answers to his vocabulary questions and asked if Driggs might have time to help him whip into shape the manuscript for his latest book, *Seventy Years of Progress*. Could he also help to find a publisher? Meeker told Driggs he had secured eight hundred subscribers and if he campaigned he could easily get five hundred more. He ended by asking, "I have wondered when the new work you are preparing comes out whether I could have a finger in the pie to push sales in Washington and canvass for 'Progress' at the same time."[7]

In March the two men corresponded about the title of the book. Driggs at first leaned toward *Pioneer Tales of the Oregon Trail* and asked if Meeker was amenable to this or if he had a better suggestion.[8] Meeker approved, but Driggs was still not happy with the title.[9] At the end of March Meeker mailed a copy of his book *The Ox Team or the Old Oregon Trail* to Driggs and this solved the title dilemma. "Many thanks for sending that volume.... It gives me the suggestion I was praying for—to make the title right. Now I feel I have it. Ox–team Days on the Oregon Trail.... The publishers are pleased with it."[10]

In May Driggs gave Meeker a progress report and promised that he would be proud of the result. He told Meeker that he had given it two solid months of work, editing along with a Miss Purcell and Mr. Greene, two of the five editors employed by the company. A Mr. Wilson was the artist at work on the drawings. Driggs told Meeker the book was not going to be like *The Busy Life* but it would contain much of what was in that volume. He promised, "Our effort has been to get the cream of the crop and to keep your style throughout."[11] While Driggs was editing and writing in New York, Meeker was doing the same in Washington. He hoped to have *Seventy Years of Progress* completed by May and suggested that Driggs submit it to the World Book Company for publishing.

Meeker then opened a new topic. He had been in correspondence with Harold Allen, son of pioneer Edward Jay Allen. He told Meeker of his father's Oregon Trail manuscript, composed around 1913, describing Allen's 1852–55 experiences on the trail and in Washington Territory. Meeker and Allen were together for a time during their respective 1852 journeys. In later life the two men reconnected and became close friends, working together on Meeker's efforts to monument the Oregon Trail.[12] On March 23 Meeker received the first installment of the manuscript that Allen titled *The Oregon Trail or Letters from the Great Plains*. Meeker told Driggs about Allen's accomplishments, praised his writing, and thought the family would be very willing to let him have the manuscript for the publication of a small number of volumes to distribute among friends. Meeker had taken a pause in working on *Progress* and had written some "scribble" he titled *High Times on the Oregon Trail*, and sent Driggs a few pages. Meeker's intention was to tell of trivial incidents that illustrated life on the trail. He also thought, combined with the Allen manuscript, he could easily fill a volume the size of a school series. "I think those letters are of high value and would be much sought after. The descendants may conclude to publish it themselves but I think they would prefer to have it jointly with some of my experiences." He asked if Driggs thought such a work would detract from *Ox-Team Days* or improve sales.[13] Driggs politely steered Meeker away from this project,[14] leaving the publication undone for nearly a century until this author and Karen Johnson undertook the work.[15]

Meeker planned to tour the state in July and August gathering subscriptions for *Progress* and was willing to hawk *Ox-Team Days* as well but noted, "However, I do not look for a very large sale here in Washington unless it varies more than I am led to believe from your request of illustrations."[16]

Seventy Years of Progress was published in Tacoma in 1921. *Ox-Team Days* was published by the World Book Company, upstate New York, in 1922. There would be one more book in Meeker's repertoire, a romantic novel about the Oregon Trail titled *Kate Mulhall*, published in 1926.

Minnie Howard

Another person who helped frame the trajectory of Meeker's final decade was his old friend, Minnie Howard. Howard, now regent of the Wyeth Chapter of the DAR in Pocatello, Idaho, invited Meeker to speak at the July 27 ceremonies celebrating the 87th anniversary of the sermon by Jason Lee at Fort Hall in 1834, the first sermon given west of the Rocky Mountains by a protestant missionary. It was an outing Meeker thoroughly enjoyed. He stayed as a guest at the Howard home, spoke to 380 teachers attending summer school, and had his expenses paid by the president of the Idaho Technical Institute. Meeker had last been to Pocatello in 1916 when, accompanied by Minnie Howard and others, he conclusively located the site of Fort Hall.

Meeker addressed the crowd at the annual celebration, which was jointly sponsored by the DAR and Rotary. All gathered around the monument placed at the site, listening to Meeker speak standing in the center of a freshly mowed field. It was a long speech giving the history of Jason Lee and Fort Hall, and of his own years-long search for its exact site. The speech was published in full by the Pocatello newspaper.[17] On Friday Meeker traced the trail from Lava Hot Springs to March Creek, and back as far as Inkom. That evening he addressed pioneers and citizens at the Riverside Hotel in Lava Hot Springs. The next day he returned to Portland by train, where his old friend George Himes offered some criticism of his effort.

> Thank you for your address at Pocatello. Considering your youth you did very well. Only one criticism I have to offer—that is, you should have your trainer show you how to put a little more punch into your voice—spiritual punch, in a certain sense. Ordinarily, I would not dare to make such a suggestion; but now that my advancing years give me some privileges that I did not used to possess, I feel free to "pint" out your oratorical shortcomings.[18]

Back home, as opportunity allowed, Meeker continued his educational campaign, speaking at venues such as Broadway and Franklin High Schools in Seattle, a history pageant at the College of Puget Sound in Tacoma, a joint gathering of Pierce and Thurston County pioneers at Point Defiance Park in Tacoma, and a clambake in Port Townsend attended by three thousand. In August and September he hit the county fair circuit visiting Enumclaw, Marysville, Burlington, Ferndale, Blaine, Mt. Vernon, Stanwood, Puyallup, and Yakima.[19]

The Borrowed Time Club, with the sole membership requirement that one had to be at least seventy years old, had a local enrollment of seventy-five, including one Ezra Meeker. Sixty club members celebrated Meeker's ninety-first birthday with a clambake at Meve's Cafeteria in Seattle.[20] He told them he looked forward to meeting again on his hundredth birthday. No doubt in preparation for this future event Meeker ordered two books from the Peebles Publishing Company—*Death Defeated* and *How to Live 100 Years*.[21] Meeker dined with his family in the evening that concluded with a public reception. The upcoming year would be pivotal in fulfilling his long-held goal of saving the Oregon Trail.

CHAPTER 16

Movie Making

In May 1922 Meeker wrote his daughter Carrie that he was planning a trip out on the Oregon Trail, perhaps going as far as Salt Lake City. That was somewhat misleading, as he intended eventually to go to New York City to sell his books, primarily *Ox-Team Days* and *Seventy Years of Progress*. Perhaps he wanted to soften the blow of yet another absence for his newly widowed daughter. Carrie's husband Eben Osborne, in whose home Ezra had lived on and off since 1905, who had handled many of his legal and financial affairs and difficulties, and bailed him out with cash on occasions too numerous to mention, died on May 7, 1922.

One factor in Meeker's decision to go out again on the trail was the creation of the Old Oregon Trail Association (OOTA) in February by Walter Meacham, secretary of the Chamber of Commerce of Baker City, Oregon. Its purpose was to preserve and mark the Oregon Trail, the very project Meeker had devoted himself to since 1905.[1] Meacham convened a gathering of northwest businessmen, tourism officials, and newspaper editors to form his association, and claimed his inspiration came from Meeker's 1906 and 1910 visits to Baker City. That may be so, but there is no evidence in the Meeker Papers of any personal contact between the two men in those years. They might have met in 1915 at the Panama-Pacific Exposition in San Francisco where they were working in their respective state buildings. On September 7, 1916, while on his Pathfinder Expedition, Meeker stopped briefly in Baker City but made no mention of Meacham.

However they might have met, the two men were working together in 1922. Over the July 4 holiday Meacham's organization produced a pioneer pageant that brought in some ten thousand visitors to Baker City. Meeker came a week early to aid Meacham in the planning and was a guest speaker at the ceremonies.[2]

Meacham's organization hired sculptor Avard Fairbanks to design a large brass medallion to mount on the markers that were to be placed along the trail from Oregon to Missouri. Three were installed—in Baker City and Seaside, Oregon, and in Vancouver, Washington. (More markers, recast from the original design, have been installed in recent years.) According to Meacham, Ezra suggested removing the wagon brake from Fairbanks' design as "in the old days they didn't use any brakes."[3]

Meeker was not present for Meacham's successful, OOTA-sponsored 1923 Blue Mountain Pageant in the hills above Pendleton that featured President Warren Harding as speaker and guest of honor. At the time Meeker was making preparations in New York City for yet another trip west over the Oregon Trail.

Mid-July of 1922 found Meeker in the Boise, Idaho, area. Here he attempted without success to have a bronze plaque installed on the Oregon Trail monument that had been erected on the state capitol grounds in 1906. Meeker hoped a plaque would solve the problem of a difficult to read inscription. Ezra also aided the local DAR chapters in their effort to place a monument at the site of the Ward Massacre.[4] He journeyed out to Sinker Creek near Murphy where in 1860 the Utter wagon train fought a running battle with an estimated twenty-five to thirty Indians. As an aside he told the local press there were 160 markers along the trail. In early August he spoke before the Rotary Club of Twin Falls.[5] By late September he was in Nebraska stirring up trouble by lobbying to have the trail on the north side of the Platte River officially labeled the Oregon Trail, in direct opposition to the state-authorized Oregon Trail Commission that had declared the route south of the Platte River to be the Oregon Trail, while the northern route would be called the Mormon or California Trail.[6]

On October 3 Meeker was in Chicago leading a parade down Michigan Boulevard. He was seated in an ox-drawn covered wagon driven by an old time "ox-skinner" named Sam Swan. Meeker's wagon was followed by Chief Big Elk of the Mohicans. Behind them came a stagecoach from the Ben Holladay line full of bullet marks from dozens of encounters with road agents, driven by eighty-one-year old E. T. Laidlaw. Behind the stagecoach came William Meehan, of Indianapolis, riding his "tall" bicycle with its carefully preserved fifty-seven-inch wheel. The parade marked the opening of the American Electric Railway Association convention.

For Meeker it was simply one more opportunity to put his message before the American people.⁷

Meeker had been intrigued with the idea of using motion pictures to advance his Oregon Trail agenda for some time.

> My first thought 16 years ago was to erect monuments on the trail to perpetuate the memory of the pioneers, but later it seemed to me that by portraying the historic facts by means of moving pictures, to be seen by millions, not only of our generation, but of the future, a greater and better result would be accomplished.⁸

In 1912 he attempted to make a feature length motion picture about the trail, but failed, primarily due to lack of finances. He estimated that to do it right might cost as much as a million dollars. Now his appetite for a motion picture project was whetted once again by news that the Universal Film Manufacturing Company was planning a serial project titled "Winners of the West," and Meeker had even written suggesting that his experience might be useful to them if they were interested.⁹ They were not.

PIONEERS OF AMERICA, INC.

By late October Meeker was residing at the Hotel Stenton in New York City, deep in planning a motion picture about the Oregon Trail.¹⁰ He had a private secretary and was selling bonds for the project at 6 percent interest. Meeker told a reporter the first person he approached purchased $500 worth. How all this came about is somewhat murky. As related earlier, Meeker was introduced at some point to George Dupont Pratt who was involved with the Camp Fire Girls organization.¹¹ It is likely this meeting took place in New York City that October. Howard Driggs said that Caspar W. Hodgson, the founder of the World Book Company, actually introduced Meeker to Pratt. However it came about, Pratt became deeply interested in Meeker's idea of a motion picture, put Ezra on a retainer, and tasked him with getting the details and costs of the project in order. It was made clear to Meeker that this had to be done before the purse strings would open for the final project. A formal agreement was drawn up between the two men. For some reason Pratt must have instructed Meeker to avoid mentioning his involvement in the film project. In his correspondence with William Bonney, Meeker stated that he was backed by a man "fifty or more times a millionaire,"

Cover of prospectus for Pioneers of America. *Author's collection*

and referred to this person as "moneyman" or the "millionaire." Only once did he supply a name. In a February 13 letter to Bonney he wrote, "Mr. Pratt (the gentleman furnishing the fund) has associated himself with me as partner in equal interest with himself and a third party by a written agreement for the preliminary work... when this is finished it is the intention of all three to reorganize when we have some data to guide what is best to be done."[12] The third party to the agreement was Allen Hendershott Eaton,[13] artist and activist, and Meeker's correspondence indicates that he was working primarily with Eaton at this point. Meeker told Bonney the final cost of the project would be in excess of $100,000 and that it would take two years to complete.

While in New York, Meeker made his working headquarters in the office of the National Highways Association. He had worked closely with various local Good Roads associations in the early 1900s, the forerunners of the National Highways Association, and thus was acquainted with many of its officers. He had corresponded with Charles Davis, the national president, in 1916 when on the Pathfinder Expedition, and again during his campaign to get the State of Washington to construct a highway over Naches Pass. A second connection with the organization was with Robert Bruce, editor of *Motor Travel* magazine, the voice of the National Highways Association. Meeker and Bruce met in 1907 when Ezra drove his ox-team into New York City. Bruce rode with Meeker one day on that occasion and invited him to lodge at his home.

In 1922 Charles Davis offered Meeker the use of the New York office as his headquarters when he was in town and, along with Howard Driggs, became a vice president in the newly formed film production company. Robert Bruce was also brought into the orbit of the film project. Early on Bruce wrote Clarence Bagley asking for a picture of the 1907 ox-team in New York City, ending by stating, "Sooner or later, the work Mr. Meeker has done will be perpetuated not only in moving pictures but in some very substantial volume; and I anticipate that your picture might form an interesting part of such a volume."[14] When Meeker left home his idea was simply to work with Howard Driggs on a plan to sell the two books. Suddenly, he was in the motion picture business.

If the partners were to make a motion picture about the Oregon Trail, it was clear that they needed a covered wagon. The wagon saga dominated the remainder of the year. Eaton wanted a covered wagon for the film identical to the one Meeker drove across the country in 1906–07. (Meeker had deeded his original covered wagon to the Washington State Historical Society in 1919, and in 1922 the wagon was on display in the Ferry Museum in Tacoma.) Pages of correspondence went back and forth between Meeker and Bonney regarding building the wagon. There were discussions about building the wagon in New York versus Washington State, and if in Washington, who should build it and how it should be shipped, whole or disassembled.

In December Meeker informed the Olympia Chamber of Commerce that he would be in New York for the next six months under an arrangement made to secure data for a serial of moving pictures on the Oregon Trail. He told Bonney it was an eight-month engagement and that he regretted he would not be home for the upcoming legislative session and urged Bonney to take on the lobbying effort for Naches Pass as well as ramrodding the construction of the covered wagon.[15]

Meeker again wrote to Bonney on January 20 with encouraging news,

> I think now that my co-workers here intend to go further than simply the preliminary work that is provided for in our present agreement. There are three of us in this moving picture project, the man with the money [Pratt], an experienced moving picture expert [Eaton], and myself, all agreed to produce a strictly historical film…My friend of wealth is eighty times a millionaire; my friend the expert is on

the staff of the Sage Foundation; neither of them is attracted to this work for gain but because of their interest in recording accurate history...there are four of us [the fourth being Robert Bruce] in perfecting some of the introductory scenes.[16]

BUILDING A COVERED WAGON

In January 1923 Meeker sent instructions to Bonney to go ahead with the construction of a covered wagon. A check for $275 soon followed. He requested that the wagon box be well caulked so it could be used as a boat in river crossings. He also asked Bonney to secure a bid for building a stagecoach similar to one in the museum. Joseph R. Turner of the West Coast Steel Company in Tacoma was put in charge of building the wagon under the watchful eye of Maurice Fitzgerald,[17] an old pioneer and friend of Meeker's. Mr. Turner also offered to build a stagecoach for $575 or two for $1,100 and he told Bonney that he wished to ship it complete rather than knocked down. Bonney reported making a visit to the West Coast Steel Company. He wrote Meeker that they "are doing good, careful work on the wagon construction." Meeker asked, "Let me hear from you as soon as it is completed and accepted by you and at the same time if it's in the way at the shop."[18]

The partners were not working in a vacuum. A motion picture by Paramount titled "Covered Wagon" was about to be released,[19] and Universal was advertising their "Oregon Trail" serial production. The partners did not see these as a threat as Meeker assured them that theirs was going to be "a strictly historical film" that was to be "accurate and to be a standard for all time."[20]

At the end of May, Meeker wrote Bonney, "I send you copy of our organization papers already signed and will be filed this week of which I will soon write you more in detail." It took some time to get the new organization under way. The partners were held up three weeks trying to get the New York Legislature to approve the use of the name "Pioneers of America" for the incorporation. Then they lost another week trying to round up a quorum so that they could do business. Five thousand copies of a prospectus were issued and approval was given for bond sales. Meeker hoped the New York people would underwrite half the sale, and he planned to make a three-month summer trip by rail and auto to arrange for the sale of the other half. He also planned to do preliminary work for the filming such as taking notes of where scenes should be shot. He hoped to be back in the northwest in time for the Oregon State Fair and the Puyallup Fair, and if the road to Paradise Valley and hotel in Mount Rainier National Park was still open, film some scenes there. Meeker was concerned that it might be impossible to film a buffalo stampede as there were so few buffalo left. He was assured that there was a herd in Salt Lake and another in Canada that would serve the purpose. He asked Bonney if he would be willing to serve as the local agent for the company and if not suggest someone that would.[21]

On Monday, June 25, at 7:45 p.m. Meeker gave a twenty-minute radio address on New York station WEAF. The radio company sent out notices of the program to five hundred newspapers, giving Meeker a potential audience of 100,000, the largest audience he ever addressed.[22] Meeker wrote, "To cap the climax the New York Times has just telephoned requesting an interview and I understand it is for a feature article for the Sunday issue. As I have said I have just gotten the machine in working order and if I was so minded I might sign this letter as President, in other words, I am the executive officer."[23]

On the Trail Again

On July 18, 1923, Meeker left New York City for yet another trip over the Oregon Trail from Kansas City to Portland. His plan was to survey potential filming locations and work out the logistics for producing the movie.[24] The prospectus asked that those interested in helping with the preliminary survey get in contact with Ezra Meeker at the New York headquarters. Despite receiving a blow Meeker carried on. George Pratt had returned from Europe and stated that he was not willing at that time to commit to a final project, although the preliminary work could continue.[25]

As scheduled, Meeker was in Kansas City at the end of July, arriving by rail from Chicago in a Pullman car. He had three locations in mind that he wanted to scout, Liberty, Missouri; the Fitzhugh Mill in Dallas, Kansas; and Elm Grove, Kansas. Liberty was important, as it was from here that missionaries Marcus Whitman and Henry Spalding, along with their wives, began their 1836 journey west. Narcissa Whitman and Eliza Spalding were the first white women to make this overland trek. Meeker spoke to a *Kansas City Star* reporter about the large wagon train that formed in 1843 at Fitzhugh Mill and Elm

AT THE
Starr King Hall
CORNER CASTRO and 14th STREET
Thursday Sept. 10, At 8 P. M.

EZRA MEEKER

Will Deliver His Lecture

THE
"OLD BOY SCOUT"

Illustrated by

STEREOPTICON VIEWS

And Three Reels of

MOVING PICTURES

EZRA MEEKER

He will relate his sixty two years experience of frontier life; incidents of his three trips across the plains, and of the Indian wars of the Northwest in which he was involved, together with his life among the Indians in times of peace

Admission

Children, 10c   Adults, 25c

Boy Scouts and Camp Fire Girls Free

Keystone Printing Company, Oakland

HIGH GRADE BUSINESS CARDS PER 1000 $2.00
500 $1.25   100 50 CENTS
PRICES ON ALL OTHER WORK IN PROPORTION

568 Ninth Street   Phone Oakland 1859

This Oakland, California, dodger advertising a September 10, 1914, show illustrates Meeker's use of motion pictures nearly a decade before the incorporation of Pioneers of America.

Grove consisting of eight hundred pioneers with 125 wagons and five thousand head of stock. It was the first wagon train bound for Oregon and the Pacific Coast. Meeker said, "The camp and start of this expedition will be one of the features of the completed film."[26]

Allen Eaton told Robert Bruce in a telephone conversation in late August that Meeker was about to enter Wyoming and was making fair progress. Bruce was concerned that Meeker's job required "all his energy and persistence" and wondered if, perhaps, it should be handled by a younger man. He also wrote, "I am afraid that the issue is more or less in doubt," confirming that Pratt was unwilling as yet to commit to actually making the motion picture.[27]

Meeker's intention was to arrive at Portland in mid-September, spend some time in Oregon and Washington and return to New York in November. His Oregon sojourn included a visit to Eugene, Oregon, where Avard Fairbanks, professor of fine arts, sculpted a likeness of the pioneer.[28]

Cheating Death Again

Meeker did not return to New York that year as planned. On October 17 his half-brother Aaron Meeker died in Bellingham at age fifty-eight. The funeral and burial was in Sumner, Washington. Ezra attended and caught a severe cold or "grippe" as a newspaper called it.[29] Upon returning to Seattle he consulted his grandson, Dr. Charles Templeton, who put him to bed in his home in order to better monitor his grandfather's condition. Ezra was running a fever of 101 degrees and had a pulse of 110. Templeton declared his condition critical, stating, "Illness of any nature is a very serious matter to a man of his age. Mr. Meeker is very, very weak."[30] The next morning Ezra's condition improved markedly. His fever broke and his pulse returned to normal. He was even able to sit up and eat. Templeton stated his grandfather would recover and was likely to be back to normal by the end of the week. On December 3 Meeker gave Driggs a progress report on his health. "My health has returned to me and I have gained back nearly half the weight I lost (9 lbs.) while sick, and my strength is returning, though slow."

Following his illness Meeker remained in Seattle. He turned over his voting authority in Pioneers of America to Howard Driggs.[31] He also told Driggs that he thought it would be necessary to establish a branch office in the West and place western directors, particularly the local timber barons, on the board before he could make any headway in securing funds to finance the company. He asked Driggs to call a meeting to consider the request. "I want now to establish the office in Seattle. The lumber interest in Washington and Oregon is tremendous and very prosperous... There are hundreds of millions invested in the lumber industry and I hope if I once get a good start we can secure enough funds. Anyway we can not know without we try and I mean to make a strenuous try."[32]

In the same letter Meeker noted that Robert Bruce had written him that the board of directors were of the opinion that they should give Meeker 10 percent of the sale of bonds as compensation for his work. Ezra felt this was a very inappropriate use of funds, as the sale of bonds was intended for the making of the movie. He replied in very firm language that he could not agree to such an arrangement and asked Driggs to endeavor to rescind any such action if a formal vote had been taken. He did feel he was entitled to some compensation, but this was not the way to go about it. Eaton and Bruce met on December 6 and, adhering to Meeker's concern, agreed to pay him $500, or $125 per month, in consideration for his four months service to the corporation before his illness took him off the job. Bruce said the balance in the treasury was very small with an outstanding bill to be paid and that they would be unwilling to pay more until the treasury was replenished.[33] Meeker wrote in the margin of that letter that he made no response effectively accepting the decision. He also noted that he was back to good health and felt fine. Bruce acknowledged the reception of a seventy-five page synopsis submitted by Meeker and Fitzgerald of proposed scenes to be filmed and noted that it only went to 1843 and did not cover Meeker's 1852 journey. He felt that *Ox-Team Days* could serve as a source for that part of the story.

On December 26 Howard Driggs sent Meeker birthday greetings, congratulated him on his recovered health, suggested he might get his wish to live to one hundred, and heaped praise on him for all he had accomplished in saving the Oregon Trail. There was no mention of filmmaking in the letter. Pratt pulled the plug on further financing at this point but Meeker was not willing to surrender.

PULLING THE PLUG

After recovering from his illness Meeker, now ninety-four years old, made his way to Los Angeles in mid-January and set up shop in the

New Hotel Rosslyn on the corner of 5th and Main. He wasn't about to give up on his motion picture. Six weeks later he reported to Driggs that he had not made much headway but was optimistic. He had learned in Hollywood that step one in the motion picture business was making a movie. The more difficult step two was getting it shown. This was a surprise to Meeker who assumed the hard part was making the movie. He also learned that the prospect of selling bonds to finance the motion picture was dim.

In mid-March Meeker had concluded that it was time to make a decision. He wrote Eaton saying they needed to hire a director and start filming or abandon the project completely. He was certain he could arrange financing for the film and had found a director who would do the job for $170 a month. Meeker estimated it would take three months. He had spent $700 in California and was about out of funds. He needed at least $500 more from New York. He wrote, "I must have help or quit."[34] Driggs suggested that Meeker go back to the idea of making short serials on various episodes of the pioneer experience, noting that the educators he had been working with preferred short films to long ones as they were easier to work into lesson plans. And he wasn't enthusiastic about the prospect of more money coming from New York.[35] Meeker finally accepted that the New York money had dried up, but he was not yet willing to quit. His last option for securing local funding by opening a branch of the corporation in Seattle and expanding the board to give the Seattle interests a voting majority came to nothing.[36] Driggs again encouraged Meeker to go back to the idea of multiple serial type episodes rather than one grand picture encompassing the entire story of westward movement. He felt this was the best and only feasible way to proceed.[37] In the end, the project went into remission but Meeker did not let go.

The 1923 Meeker Covered Wagon

And what became of the movie wagon that was built in early 1923? It was intended to be an exact duplicate of Meeker's original expedition wagon from 1906 that resided at the time in a glass case in the Ferry Museum in Tacoma. It was as authentically reproduced as possible, with Mr. Turner's workers going into the glass case housing the wagon to take measurements. They even tried to match the finish. The plan was to ship the wagon to New York when it was completed, and use it in the preliminary filming and advertising for the Oregon Trail movie. A decision on how to ship it east had not been made when Mr. Pratt stopped financing the project.

It seems the wagon never made it out of the northwest. When Meeker was in Washington in 1926 the wagon was at the dedication of his statue in Puyallup's Pioneer Park. In November 1926 a photograph was taken of him selling Oregon Trail memorial coins from the wagon in Tacoma. That photograph showed a large, distinct crack in the footboard that has been instrumental in tracing the subsequent history of the wagon. Somehow, it made its way to Portland, where, perhaps, it was used as a prop for the selling of coins in that city. A May 25, 1927, letter from Meeker to the Ford Motor Company in Portland, Oregon, stated that he had asked his grandson, Frank Templeton, to make arrangements to store the wagon elsewhere, but that Frank had not responded. Meeker asked the Ford Company to telephone Templeton, and if he declined to get involved could they find a place to store the wagon at a reasonable cost. Instead the company billed Meeker fifteen dollars for storage. Ezra stated "[T]he wagon does not belong to me," and declined to pay the bill. He told Mr. Henderson that he had written George Himes to see if he could get one of the pioneers to take care of the wagon until some disposition was made. Himes eventually took charge, and Meeker paid the Ford Motor Company twelve dollars for their trouble. The wagon found a home in a storage facility at East 11th and Division Streets in Portland.

Fittingly, the wagon was called into service for Meeker's funeral. On December 5, 1928, the *Seattle Post-Intelligencer* stated that William Bonney brought the wagon up from Portland for that purpose. A photograph taken of the wagon as it was carrying flowers to Meeker lying in state at a local funeral home, showed a large, distinct crack in the footboard, identical to the 1926 photograph. The wagon appeared again in local newspaper stories being driven by Charles Ross and Frank Spinning in a parade at the Southwest Pioneers' Association picnic in Centralia in August 1931 where it won first prize.[38] This photograph also showed the distinctive crack.

The Puyallup Historical Society at the Meeker Mansion (previously the Ezra Meeker Historical Society) has a wagon they use for public appearances that matches Meeker's 1906 wagon in size and construction, and has a large crack in the footboard that is identical to the crack that is shown in the 1926 photograph, the 1928 funeral photograph, and the 1931 Centralia gala. Andy Anderson, of the society, said in 2014, "'Our' wagon was acquired by the Days of Ezra Meeker Corporation, and was then stored in the field of Mr. White, outside of Orting, before it was then acquired by the Ezra Meeker Historical Society."[39] The Days of Ezra Meeker Corporation was formed on August 22, 1939. Its first festival was in September of that year. The festival program listed a street parade on September 16, 1939 that featured the Ezra Meeker covered wagon. All this suggests that the wagon remained in or around Puyallup, Washington, following Meeker's funeral.

Meeker's movie wagon at last made a trek over the Oregon Trail, in commemoration of the one-hundredth anniversary of Meeker's 1906 Old Oregon Trail Monument Expedition. The then-Ezra Meeker Historical Society, jointly with the Oregon-California Trails Association, trucked the wagon and a pair of oxen over the trail from Puyallup to St. Joseph, Missouri. At twenty stops along the way re-enactors in period attire shared Meeker's story. Ezra Meeker himself, in the person of re-enactor Ray Egan, highlighted this anniversary journey. It appears the wagon, now approaching ninety-seven years of age, is still serving a purpose of which Meeker surely would have approved.

CHAPTER 17

The World's Oldest and Youngest Aeronaut

First Flight

Meeker's first encounter with airplanes took place in 1910 at the Southern California air show where he was simply a spectator. On September 25, 1919, at age eighty-nine, he took his first flight in an airplane piloted by Percy Barnes of Seattle. The goal was to fly from Tacoma to Mount Rainier, but smoke precluded that plan. Instead they flew near the mountain and over Tacoma and Puget Sound for an hour. "It was wonderful," said Meeker who had just set an age record for flight. "I thought I might be nervous or dizzy when I got into the airplane but I wasn't the least bit bothered. Why, we flew over ground that I traveled over sixty-five years ago…. It was simply beautiful."[1]

A Personal Best

On May 27, 1920, Meeker recorded another milestone in his illustrious career. Now age ninety, he accompanied pilot Eddie Hubbard on a round trip U.S. mail flight from Seattle to Victoria, British Columbia. Meeker, with his white hair hidden beneath a leather helmet, climbed into the cockpit of Hubbard's mail plane at 6:30 a.m. and landed in Victoria forty minutes later, shattering his own 1858 record. Meeker had made his first journey to Victoria in 1858 at the height of the Fraser River gold rush. That trip, made aboard the steamboat *Constitution*, took seventeen hours and was believed to be an unbreakable record at the time. On this trip Hubbard was making a mail run to the trans-Pacific liner *Arizona Maru*. He dropped Meeker off for the day in town and returned to Seattle to pick up another load of mail for the Canadian liner *Empress of Russia*. Meeker spent the day visiting the Tolmie family. He also presented the government library in Victoria with a copy of a recently discovered manuscript that told of his early experiences in Washington. Meeker, who had forgotten about the manuscript until he found it among the effects of the late Joseph Kuhn of Port Townsend, copied it by hand for the library.[2] On the return trip, Meeker again rode in the plane as passenger, arriving shortly before 7 p.m.[3]

CLARKSTON TO SPOKANE

Meeker's 1920 day calendar listed visits to cities around the state where he was gathering material for his latest book project, *Seventy Years of Progress in Washington*. In June he ventured east of the Cascades to Clarkston where he gave a lecture in the city park to the pioneers of Asotin County. While he was speaking a plane flew overhead and dropped messages of greetings upon the crowd. Meeker turned to his host and asked if it would be possible for him to fly to his next day's engagement with the DAR in Spokane. The president of the Chamber of Commerce left the gathering and returned with word that the pilot would be glad to fly Mr. Meeker to Spokane. Thus Meeker made his third flight—duration two hours, distance ninety miles.

Meeker made several other flights around the state during his campaign to have the Naches Pass wagon road selected as a cross-mountain highway. By the fall of 1924 he was a seasoned flyer with ten flights under his belt, but without question the upcoming transcontinental jaunt that would take him to Washington, DC, and a meeting with President Calvin Coolidge would become his favorite.

Meeker's third airplane flight, Lewiston/Clarkston to Spokane. *Seventy Years of Progress*

Transcontinental Flight

Lieutenant Oakley G. Kelly of the U.S. Army Air Service was a national celebrity before he and Meeker teamed up in 1924. In 1922 Lieutenant Kelly flew from Long Island, New York, to San Diego, California—the first solo transcontinental flight. Kelly and Lieutenant John A. MacReady set the world flight endurance record when they made a non-stop transcontinental flight from New York to the Pacific Coast May 2–3, 1923. From 1924 to 1928, Lieutenant Kelly was the commander of the 321st Observation Squadron based at Pearson Field in Vancouver, Washington, where he earned a reputation as a daredevil pilot by often flying under Portland's Broadway Bridge.[4] After attempts by several other countries failed, the United States military sought to be the first to fly around the world. Four Douglas World Cruiser biplanes were specially built for the U.S. Army at Santa Monica, California. They were named *Seattle*, *Boston*, *Chicago*, and *New Orleans*. Kelly was one of the escorts to accompany the flyers from Santa Monica to Sand Point Air Field in Seattle. The contestants departed Seattle on April 6, 1924. Only the *Chicago* and *New Orleans* completed their journey, arriving back in Seattle 175 days later, on September 28.

In the summer of 1924, Meeker heard that Kelly intended to fly from Vancouver, Washington, to the Dayton, Ohio, air show in October. Meeker immediately began lobbying the War Department to go along as a passenger. As the the official representative of the Seattle Chamber of Commerce, he wanted to confer with President Coolidge about opening an airmail route from Elko, Nevada, to Pasco, Washington. Such a route would shorten the existing airmail time between New York and Seattle by forty hours, and it would make possible twenty-four hour airmail service between Seattle and San Francisco.[5] Meeker also wished to contrast the covered wagon period with the present age of the airplane. Privately, Ezra knew such a flight would generate an enormous amount of national publicity that would greatly benefit his Oregon Trail cause. Preparatory to the flight Meeker made arrangements to send telegrams at refueling and overnight stops to the Seattle and Tacoma newspapers, knowing full well that the story would go over the wires and become national news.

Meeker was at the Puyallup Fair on September 30 when he finally received special dispensation from Assistant Secretary of War Dwight F. Davis. He immediately took the night train from Seattle to

Vancouver. By 6 a.m. Meeker was sitting beside the airplane waiting for his pilot, who arrived three and a half hours later. The *Seattle Post-Intelligencer* reported, "Before hopping off, the elderly flyer, as a precaution, removed a set of false teeth from his mouth and placed them in his pocket for safekeeping. His manner was cheerful, resembling much that of a school boy off on a big lark. He talked cheerfully to spectators gathered about the field and declared he had not the slightest doubt but the flight would be crowned with success."[6] Around 10 a.m. on October 1, 1924, he was on his way in a U.S. Army De Havilland-DH 4 airplane, becoming one of the nation's first transcontinental air passengers. It was, Meeker said, his eleventh airplane trip.

The pilot and passenger circled the field once and then followed the Columbia River east through the gorge, crossed the Blue Mountains where they encountered a brief snow storm, and landed at Barker Field, Boise, Idaho, at 2:10 p.m., averaging 120 miles per hour. At 3 p.m. the duo took off for an hour and thirty minute flight to Pocatello where they spent the night after landing in an alfalfa field. In one day Meeker covered six hundred miles or one third of the Oregon Trail. In a telegram to the *Post-Intelligencer* Meeker wrote, "Lost my $8 hat and here in Pocatello bareheaded: no matter. I have plenty of hair on the top of my head. Trip 'bully,' as Roosevelt would say."[7]

On day two the DH-4 airplane reached North Platte, Nebraska, at 4 p.m. after gasoline stops at Rock Springs, then Cheyenne, Wyoming. Meeker's telegram said his only complaint was that his nose would get cold as they were flying at 8,000-feet elevation much of the time. The airplane also hit an air pocket dropping one hundred feet. Meeker was asleep at the time but the jar awoke him with a start.[8] He commented, "I reckon if it hadn't been for the strap around my waist, I'd still be floating around in space trying to find the up and down of things."[9] It was another 600-mile day.

On day three they ran into strong headwinds. Meeker telegraphed, "Our ship would suddenly drop a few feet and then as suddenly rise again then gracefully roll to one side, then the other as if through veritable air waves, all the while, speeding through the air a hundred miles an hour."[10] During a fuel stop at Jarvis Offutt Field at Fort Crook, Omaha, Nebraska, Meeker, after shedding his helmet and aviator gear, was greeted by Major L. S. Churchill, air officer of the Seventh Corps, and Colonel C. C. Kinney, commandant at Fort Crook. In thirteen hours and thirty minutes of flying time Meeker covered the distance that took him five months by ox-team in 1852. The Omaha press reported that Meeker jumped from the plane "as spry as a man of 40." Kelly noted that Meeker's first words in the morning were "Let Her Rip." Said Kelly, "In the air he is continuously asking me for more speed. When night comes and I start to land, he begs me to go just a few miles further. Only when I explain that we are out of gasoline is he content to stop."[11]

The *Omaha World Herald* quoted Kelly, "There stands the world's oldest and youngest aeronaut!... He's all but worn me out with his pep. When it's time to stop he wants to go farther. He's the original 'let's go' man of the country!" Meeker noted that back in 1852 he crossed the Missouri River with an ox-team and that it took him four months and twenty-seven days to cross the continent. "Which do I like the best? Why an ox team of course," he said. "You can't change an old fellow's tastes this late in the game. But I like planes and I've had a great trip so far."[12]

He went on to tell the Omaha reporter,

> My principal interest in life today is to work for good roads, and put over this Oregon trail idea.... When Roosevelt was President I got an act of Congress through to improve the Oregon trail, now I'm going to see President Coolidge and see if he won't get busy and help make it a great paved highway.[13]

They hoped to reach Indianapolis but only made it as far as Rantoul, Illinois, where they landed at Chanute Field at 4 p.m. and spent the night.

Day four the flyers were off at 6 a.m. for the final leg of their journey to Wilbur Wright Field in Dayton. They landed at 10 a.m. October 4 and were met by Wilbur B. Moore, secretary of the Dayton Chamber of Commerce. Meeker and Kelly were placed on an elevated seat in an open automobile and paraded in front of the mile-long grandstand holding some 100,000 spectators at the air show. The audience began to cheer and the ladies waved their handkerchiefs. At first Meeker sat quietly, simply looking around. Then Kelly prodded him to respond, saying the audience was cheering for him, so Ezra began to wave. After a slow mile-long drive in front of the grandstand Meeker was escorted to the box occupied by Orville Wright, who commented that he thought his father was quite amazing when flying at age eighty-three. He told Meeker, "I guess you've got it on him." Meeker responded by saying, "I'm letting nothing pass by. I've watched the world's progress too long not to want to ride in a plane. You'd be surprised at the difference between riding in a prairie schooner and in an airplane."[14] After a few hours watching the air show, Meeker was placed in front of a microphone where he gave an interview for the local radio station, which Kelly described as "excellent."

On October 6 Kelly and Meeker did a four-hour jump from Dayton to Washington, DC, where the pilot turned Ezra over to Harold Allen, son of 1852 pioneer Edward Jay Allen. Meeker delivered the missive from the Seattle Chamber of Commerce to President Coolidge and urged him to sign on to his Oregon Trail projects.[15] Business concluded, Meeker took the train to New York City, touched base with his Pioneers of America crew, and returned home to Seattle.

On January 18, 1925, Meeker gave a talk on KJR radio in Seattle about his recent flight across the continent. Sportscaster Royal Brougham, who came on the air after Meeker, said it was a pleasure to follow "such a grand old sportsman as Mr. Meeker, who was a great ox-cart racer in his day.... He holds the record for galloping across the plains...in six months, two weeks and nine hours."[16] The next stop for Meeker would be the circus.

Meeker and President Coolidge on the White House lawn, October 7, 1924. *Library of Congress LC-DIG-npcc-12375*

CHAPTER 18

"No Half Hours to Spare"

On February 19, 1925, the *Tacoma News Tribune* related the following: "Ezra Meeker, seeking a friend in the offices of a Tacoma newspaper, was asked to wait for the friend's return, which would be in half an hour. 'Good gracious,' was the answer, 'I am ninety-five years old, and I have no half-hours to spare.'" Indeed, Meeker was a busy as ever that year, first with a Wild West show, then with the publication of his first novel and eleventh book.

THE 101 RANCH WILD WEST SHOW

In 1905 the Miller brothers, Joe, George, and Zach, sponsored a program on their sprawling Oklahoma 101 Ranch that featured some two hundred cowboys and the Indian chief Geronimo, all doing various stunts. It drew an audience of 65,000 and was such a success that the brothers decided to go into show business forming the 101 Ranch Real Wild West Show, a production along the lines pioneered by Buffalo Bill Cody. By 1914 the company was performing internationally. In 1925 they did thirty-three performances in front of an estimated 700,000 people including King George V and Queen Mary of England.

The 1925 season opened for business on April 22, in Oklahoma City, with Ezra Meeker a member of the cast. One account said the ninety-four-year-old Meeker insisted on a ten-year contract, explaining he wanted "a permanent job, not one that would play out in two or three years."[1] He signed a contract in February and agreed to appear at the Oklahoma City debut and at every show thereafter for eight months, a little short of the ten years he supposedly requested. The contract stated that Meeker was to drive an ox team similar to that in which he crossed the plains in 1852 and participate in some dramatic scenes.[2] Meeker saw it as a way to drum up interest in his motion picture project. "I consider this show business the initial step in getting a true picture of the Oregon Trail produced," he said. "I hope in this way to get the attention of the big producers."³ The fact that the Miller brothers had entered into the field of movie making with their 101 Ranch Bison Film Company, and had made some of the nation's first western films, likely spurred Meeker's interest. To that end, Meeker spent his time in Seattle over the winter engaged in a new writing project. The newspapers noted that when he boarded the train heading east he carried with him the manuscript of his novel, "Romance of the Oregon Trail," that he hoped would become the basis of a motion picture.

Meeker, as usual, had another project percolating. He asked Zach Miller, the manager of the show, for permission to call on local DAR regents in towns where they were performing. His purpose was to enlist DAR support for his upcoming request to Congress to mint commemorative fifty-cent pieces which would be sold for a dollar, the profits used to purchase and place markers along the Oregon Trail.⁴ Miller agreed as long as Meeker was present for the two daily performances called for in his contract.

On September 6 the show's performers were resting in Davenport, Iowa, prior to a Labor Day performance. Clara Meeker Dice, who lived in Cedar Rapids, Iowa, had been corresponding with Ezra trying to establish her family genealogy. She met Meeker in Davenport on his day off and he gave her a signed copy of *A Busy Life*, telling her he believed they came from the same stock.

On September 8 the Wild West Show was to perform in Cedar Rapids, and Mrs. Dice invited Meeker and his traveling companion Tex Cooper, who did the announcing for the show, to be her guests at her home. It was pouring rain when Mrs. Dice and her son Raymond met the train. The guests were brought to the Dice home where they ate a hurried breakfast. Then mother and son drove Meeker to the home of Claire Skinner Johnson, the regent of the Ashley Chapter of the local DAR where Meeker lobbied her to sign on to the coin project. Their next stop was at the local bank where Ezra deposited his weekly salary to be credited to his account in the New York Stock Exchange Bank. In between stops Meeker took time out to participate in a street parade, a staple of the Wild West show. The parade was held rain or shine in every town the show visited. The children loved it and it was the only part of the show that many of the little ones ever saw.

Next, it was back to the Dice home for lunch. In the afternoon Meeker had a performance to do as well as one in the evening, both of which Mrs. Dice attended. In her memoir, *The Prairie Years*, she wrote:

> Most of the 101 Ranch Wild West Circus was rodeo in nature, but Ezra Meeker's part was pageantry, as he depicted with his ox team and covered wagon the trek of the early pioneers ever westward toward the setting sun. It seemed very real indeed, as Uncle Ezra rode out across the plain, his long white hair and flowing white beard blowing in the wind. And then the Indians came—Navajos, Apaches and Sioux—all in their war-paint fury, whooping and riding round and round the pioneer, with arrows flying in every direction. Somehow Uncle Ezra always managed to escape their tomahawks and spears and, by the merest chance, reached the shelter of the fort just in time...he had reached the city of refuge, and so all ended well! We all enjoyed it.[5]

At the conclusion of the evening performance Mrs. Dice and her son accompanied Ezra and Tex Cooper to the train to bid them goodbye. Meeker performed with the Wild West show at least through September 1925. However, he was back in New York in December for the climax of his grand quest to save the Oregon Trail.

Ezra Meeker and Tex Cooper of the 101 Ranch in 1925. *Author's collection*

Kate Mulhall: A Romance of the Oregon Trail

Meeker's one and only effort at writing fiction culminated in the manuscript that he brought with him to Oklahoma in 1925. It was intended to be the basis for the long-planned motion picture series on the history of the trail. Originally titled "Romance of the Oregon Trail," it eventually became *Kate Mulhall: A Romance of the Oregon Trail*. Clarence Bagley did some early proofreading of the manuscript but eventually Robert Bruce became the primary editor. The Oregon Trail Memorial Association, an organization founded by Meeker and others in 1926, did most of the marketing and paid the publishing costs. The book went to the printer in April 1926. A problem with the ink developed when printing the jacket, which delayed the release for a time.[6] The books were sold at two dollars a copy and 40 percent of the profit went to the association as a donation from Meeker. By October 1926 sales began to drop off but Bruce wrote Clarence Bagley that well over half the costs had been recouped.[7] Bagley replied that he had warned the New York people that sales of the book would be slow as the people in the eastern half of the country had little interest in the Oregon Trail.[8] By mid-October Bruce told Bagley that only $840 was due to the printer and that Meeker's sales on his trip west were not yet accounted for and two non-related items were included in the printer's bill. Furthermore, the office had one thousand paid-for copies available to be sold. He assured Bagley that costs of the book would be totally paid off in a few months.[9] On October 9 Bruce told William Bonney the same thing. *Kate Mulhall* eventually realized a profit, but the movie it was designed for was never produced.

The Perils of Celebrity

Perhaps the strangest exchange of letters in the Meeker papers occurred about this time, attesting to the downside of celebrity. A reporter contacted George Himes in Portland with a series of questions about Meeker. Himes wrote Bonney about it. The questions were straight out of tabloid journalism. The reporter wanted to know if Mrs. Meeker died in an insane asylum and if she was insane before she died. The reporter also asked if Meeker had then married a French flapper, when and where the marriage took place, her name, age, and history, why Meeker had married her, and when they divorced. He finished by asking Himes about Meeker's integrity suggesting that he could not be trusted in the commemorative coin matter.[10] Bonney replied that he attended Eliza Jane Meeker's funeral, that she suffered from dementia, had for a time been in a sanitarium, but never in an insane asylum. He continued, stating that Meeker never remarried. As to his integrity, Bonney recited the story of the Puyallup bank failure in the 1890s when Meeker repaid the depositors partly with his own funds.[11]

CHAPTER 19

The Oregon Trail Memorial Association

BEGINNINGS

As 1925 dawned, a group of Pocatello, Idaho, citizens led by Dr. Minnie Howard were working to put up a monument at the site of old Fort Hall. Because dams were about to be built on the Snake River and people assumed the site would be flooded, their proposal called for a one-hundred-foot-shaft equipped with a beacon light that could be seen for twenty-five miles. The monument would be built on an island in the lake that would be created by the dam. Drafts of plans for the proposed monument were made by the New York City architect Palmer Rogers.[1] To fund it, Mabel Murphy of Pocatello came up with the idea of a commemorative coin. She noted that such a coin was struck for the Stone Mountain, Georgia, project,[2] and a similar Oregon Trail coin to fund the Fort Hall monument would be a perfect solution. The *Idaho State Journal* took up Murphy's idea in an editorial published on April 16, 1925, and proposed that the coins be made from Idaho silver. Of course, news of the project soon reached Meeker, and in the April 1925 issue of *Motor Travel* magazine he gave Minnie Howard and crew some publicity. He authored an article about Fort Hall in which he gave its history and reiterated his view that it was the most important site along the entire length of the trail. He discussed the new monument proposal and used Palmer Rogers' drafts to illustrate the article.

Meeker was working under contract for the 101 Ranch, but in his spare time he promoted the coin idea with DAR chapters around the country. Meanwhile, back in Idaho the nature of the coin was taking shape. According to coin historian Gary M. Greenbaum, Pocatello insurance salesman F. C. McGowan took a Stone Mountain half dollar from his pocket and said, "Yes. Coinage. Like this!" He suggested

195

Representative Addison T. Smith of Idaho (left), Ezra Meeker, and David G. Wylie of New York, Secretary of the Oregon Trail Association (right), display Fort Hall monument plans. *Library of Congress, LC-DIG-npcc -15576*

that the coin depict the proposed Fort Hall obelisk on one side and, "On the reverse, I would have a representation accurately pictured of the covered wagon, driven by Ezra Meeker when he came across the plains." The Idaho people began the process of setting up a nonprofit to receive and sell the coins and purchase the land where the monument would be located.[3]

Meeker's ambition was much broader. He quickly realized that Mrs. Murphy's coin idea had merit beyond just funding the Fort Hall monument. He envisioned a national organization into which he would absorb the Idaho group, albeit with an important leadership role. Congress could authorize a commemorative coin, to be sold by a national organization created to market it, the proceeds then used to erect hundreds of monuments along the Oregon Trail. Representatives and senators from Washington State had been introducing and promoting bills in Congress for Meeker for years, beginning with Congressman William E. Humphrey in 1907. There would be no problem getting such a bill in front of Congress. Indeed a bill that simply designated the route of the Oregon Trail as of national significance had just been introduced in the House and Senate. The House Committee on Roads held hearings on the measure beginning January 23, 1925, and continuing on February 13, 19, and 21. While producing nothing more concrete than a 205-page transcript, the hearings educated members of this Congress about the trail, its history, and its role in anchoring the western states to the rest of the nation. The stage was now set for the final chapter in Meeker's grand quest.[4]

This new organization would then take up the goals that had eluded the Pioneers of America film corporation in 1923. Howard Driggs wrote,

> Mr. Meeker was confident every American citizen would be eager to possess such a memento and at the same time to help forward a great cause by paying a dollar for the coin. The coin, he was sure, would bring millions of patriotic citizens into active participation in the work of the Association and would provide ample funds for this work. With these funds, Mr. Meeker proudly asserted, the old trail could be marked from end to end; the story could be portrayed in motion pictures, true to the verities of the great romance; books could be created to preserve the epic; the historic shrines along the way could be restored; and a stately memorial could be erected in the national capital to the stalwart winners of the West.[5]

In December 1925 Meeker was in New York to form a core organization,[6] and on January 9, 1926, the Oregon Trail Memorial Association (OTMA) was born.[7] Along with Meeker the board of directors consisted of Charles Davis, president of the National Highway Association; Palmer Rogers, the architect of the Fort Hall monument; Caspar W. Hodgson, president of the World Book Company who introduced Meeker to George Dupont Pratt in 1923; Chauncey M. Depew, president of the New York Central Railroad; Edmund Seymour, a New York bond broker who got his start in Tacoma, Washington, in 1889; Guthrie Y. Barber, a New York banker, wholesale dealer in women's clothing, and collector of western Americana; Martin S. Garretson, secretary of the American Bison Society and curator at the Bronx Zoo; and John Abner Marquis, secretary of the Presbyterian Board of U.S.A. Home Missions.[8] Just as in 1922, Davis provided office space in the National Highways Association building in lower Manhattan for the new organization. Meeker, of course, became its first president.

The officers quickly began recruiting members. They went after the wealthy and famous for both honorary and official memberships. Among them were President Calvin Coolidge, inventor Thomas A. Edison, and Mrs. Edward Harriman, heir to a $600,000,000 estate consisting of the ownership of the Union Pacific and Southern Pacific Railroads and the Wells Fargo Company.[9] By the end of January they had enlisted twenty-one life members and thirty-one annual members and were working on Congress to pass the coin bill.[10]

Success in Congress

On January 26, 1926, the day OTMA was incorporated, U.S. Representative John Miller of Washington introduced H.R. 8306 authorizing the sale of six million coins to be sold exclusively to the OTMA at par.[11] Howard Driggs said the bill was sponsored by Addison Smith, Representative from Idaho, and Senator Wesley Jones of Washington.[12] Either way by the end of January 1926 the coin bill was in front of Congress.

Robert Bruce was working on a map of the Oregon Trail for inclusion in *Kate Mulhall*, which resulted in a good deal of correspondence with William Bonney and Clarence Bagley. In one of these letters he wrote, "Mr. Meeker is in Washington where he appeared on Tuesday [March 2] before House Committee on Coinage, Weights and Measures on behalf of his [coin] bill."[13] The committee passed the bill and sent it to the full House for consideration where it passed on April 5 without opposition. Passage in the Senate was more difficult. The bill was sent to the Senate Banking Committee where it was held, primarily due to a written protest from Treasury Secretary Andrew W. Mellon. The Secretary argued that only coins of "national importance to all people" should be struck and suggested that a commemorative medal would be more appropriate than a coin. Meeker had testified many times over the years in front of various congressional committees. He had managed in the past to get his various Oregon Trail bills out of one house or the other, but never both. Meeker was at his eloquent best on April 27 when he testified in front of the Senate Banking Committee. He testified that the coin bill and Oregon Trail were of "national importance." In a letter to Allen Eaton, Robert Bruce said that Meeker pointed out to the committee, the fact that the National Road and the Oregon Trail not only link up the great transcontinental route, but that he had personally talked with President Roosevelt about the survey, location and improvement of this as a National Post and Military Highway. He was able to point to the passage of that bill through the Senate in 1916, as recorded in his *Busy Life of 85 Years*, though it was lost in the House.[14]

Mellon responded with a letter to Meeker in which he reaffirmed his position on the coin bill. Unbowed, Meeker lobbied each of the senators in states along the National Road and the Oregon Trail asking them to stand behind the bill "as a matter of great national public interest." He contacted the governors of the trail states and urged them to telegraph their senators, asking their support for the measure. Bruce wrote that Meeker was working sixteen hours a day on the bill. Just before he departed to attend the fateful vote Ezra told Bruce that, "if necessary, early Sunday morning he would go to Senator Borah's residence and wait on the door-step until the Senator from Idaho should agree to support the Bill."[15] Still he needed some help, which came in the form of lobbying from Minnie Howard and Idaho Senator Frank Gooding.

On the eve of the committee vote Bruce wrote Clarence Bagley of Seattle, "Mr. Meeker left last night on his 3rd or 4th trip to Washington in behalf of his Bill, and to-morrow is the fateful day. I fear a reaction if it fails, as he has pursued it with almost fanatical zeal for several weeks, without respite."[16]

The banking committee reported the bill to the Senate without recommendation. On May 10, 1926, with Ezra Meeker watching from the gallery, the United States Senate set in motion a goal that Meeker had been working toward for two decades—getting the United States government involved in saving the Oregon Trail. The vote was unanimous. President Coolidge signed Public Law 325 on May 17, 1926, and gave the signing pen to the OTMA. Today the pen resides in the Howard Driggs collection at Southern Utah University.

The next step was to choose the design. Two days after the bill signing Robert Bruce presented a preliminary design to the OTMA executive board. One side featured a covered wagon, the other a map of the Oregon Trail.[17] Minnie Howard weighed in with the view that simply depicting an oxen and wagon would be unlovely unless the artist can demonstrate "that this ox-drawn wagon conveyed the most precious human freight."[18] Meeker and Palmer Rogers met with Gutzon

The 1926 Oregon Trail Memorial coin, designed by James Earl Fraser and Laura Gardin Fraser. *D. Larsen photograph*

Borglum, the designer of the Stone Mountain coin, and found the asking price too high and the time frame too long. After finding two other artists unsatisfactory, Meeker visited the American Numismatic Society, which gave him six other names to contact, among which were a husband and wife team, James Earl Fraser and Laura Gardin Fraser. Laura Fraser designed the 1921 Alabama Centennial half dollar, the Grant memorial fifty-cent piece, and the Fort Vancouver Centennial fifty-cent piece, among others. Best known for his sculpture "The End of the Trail," James Fraser designed the Buffalo nickel and portrait busts of Alexander Hamilton and Theodore Roosevelt.[19] With a nudge from Minnie Howard, the OTMA chose the well-qualified pair. The negotiated fee was two thousand dollars. A final decision on the design needed to be made before the Frasers could go to work.

Oregon Trail Memorial Association officers suggested that the coin depict seventy-five-year-old Ezra Meeker leading the wagon. Minnie Howard telegrammed New York that it should be twenty-two-year-old Meeker leading his family west to capture the true feeling of the westward migration. Again Howard prevailed. The design was ready by August 1 and submitted to the U.S. Fine Arts Commission for approval. The director of the commission was so impressed that he had a framed photograph of the design placed in the commission rooms.[20] Approval was given and minting was scheduled for September.

On the Trail One Last Time

Meeker used the interim to trek one more time over the Oregon Trail. It would be his final trip. He was ninety-five years old and had much still to do. The OTMA secured a vehicle for him, a more modern touring car than the old Pathfinder, equipped with a kitchen and bed. Daniel R. Maue, an ex-serviceman and a newspaper writer, signed on as Meeker's chauffeur, publicity man, and personal attendant. The two men planned to stop at cities along the trail, contacting local banks, chambers of commerce, and other organizations for pledges to purchase the coins when they were minted.[21] Meeker's sendoff included a toast at city hall before departure. He bounded up the steps two at a time to obtain the official good-bye from the city aldermen. They inquired if there was anything they could do anything for him. Meeker replied, "Do I look like I need anything done for me? I'm ninety-six but as pert as in the days of '52."[22]

Ezra Meeker and his driver, Daniel R. Maue, on Meeker's last trip over the Oregon Trail in the summer of 1926. *Author's collection*

Meeker and Maue left New York City on July 15 and made stops at Buffalo, New York; Cleveland, Ohio; Indianapolis and Burlington, Indiana; and Des Moines, Iowa.[23] By July 24 they were in Cedar Rapids, Iowa, but the pace exhausted Meeker. After covering 245 miles he was too tired to talk to the reporters.[24] After a brief rest, a revived Ezra spent part of the day swapping stories with ninety-two-year-old "Uncle Henry" de Long. The topic of a footrace between the two nonagenarians came up. Meeker offered to wager one hundred dollars, the winner donating it to the trail-marking fund. Uncle Henry declined saying he wasn't willing to trust his legs for that much money.[25] Later during his stay in Cedar Rapids Meeker made arrangements with local banks and merchants for selling coins.

On July 30 Meeker participated in the Gering, Nebraska, Oregon Trails Day parade. He donned an old settlers badge and led the parade down Main Street.[26] While in Gering he secured orders from the local banks to purchase one thousand coins. Meeker had become the personal face of the Oregon Trail for many Americans. He was approached by a man who lived near Scottsbluff and whose farm abutted the Oregon Trail. During introductions the farmer noted, "I met Mr. Meeker the first time he was through here." "Oh, go on," answered Mr. Meeker. "I have a boy as old as you are. How old are you?" The farmer was, in fact, referring to Meeker's visit of 1906 and not his first crossing of the plains in 1852. The boy Meeker referred to (his son Marion) was then just a babe in arms.²⁷

That evening Meeker took the Burlington train to Casper, Wyoming, with Maue following in the car. In less than two weeks the pair were in Oregon where they stopped at Emigrant Springs in the Blue Mountains for a meeting with Walter Meacham. Ezra was delighted to find some tracks of the old trail. A photograph shows Meeker drinking water from the springs with Meacham looking on.²⁸ They continued to Walla Walla, then Pendleton. Meeker told the press since he left New York July 15 he had personally arranged for the distribution of 25,500 coins which where scheduled for release by the Philadelphia mint on September 1.²⁹ Walla Walla banks would buy 10,000, La Grande 2,000, and Baker City 4,000. Business concluded, he moved on to The

Ezra Meeker at the September 14, 1926, dedication of his statue in Pioneer Park in Puyallup, Washington, with his descendants and in-laws. His daughter Ella is standing to Meeker's left and daughter Carrie is to his right. *Courtesy of Puyallup Historical Society at the Meeker Mansion, 2014.009.001*

Dalles.³⁰ The climax of Meeker's trip through Oregon was the August 19 Pioneer Parade in Eugene. Nearly 1,500 persons participated in the pageant in front of 6,000 spectators. Meeker had some competition for the starring role. E. J. McClanahan, who drove the Portland to San Francisco stage in 1859, drove a replica stagecoach in the parade. The pageant's theme was depicting progress in transportation.³¹

In early September Meeker was in Puyallup for the dedication of his statue in Pioneer Park. The local citizens had been at work since February raising funds and doing the preparations.³² Some thirty of his descendants and relatives attended the unveiling. Meeker gave a speech to five thousand attendees in which he celebrated the pioneers and talked of the cabin where he first lived in Puyallup.³³ The Puyallup *Valley Tribune* reported:

> When Mr. Meeker arrived in Puyallup a few days ago all his friends supposed that he would spend his time resting up after his long trip and wait around for the unveiling next Tuesday [September 14].... Instead, every extra moment has been spent in Seattle, Kent and nearby towns, as well as Puyallup, placing the Memorial coins in banks, ready for sale. Friday he will go to Olympia to place more coins.³⁴

The next day he was back in Seattle meeting with bankers. Meeker told the Seattle businessmen that the bulk of the first issue of coins had already been subscribed to by banks around the country. He said the Portland banks recently authorized a $50,000 investment in them, and elsewhere some $75,000 worth have been disposed of, leaving but a few more to be sold before actual coinage would proceed.³⁵ On October 4 he was in Spokane where he told the Chamber of Commerce that ten banks and three stores in Seattle had agreed to sell the coins.³⁶

On November 15 Meeker led a parade through downtown Seattle in his replica wagon, pulled by a tractor instead of oxen, to generate publicity for the sale of the memorial half-dollar.³⁷ At the scheduled start time in front of city hall Seattle skies greeted Meeker with a deluge. He acknowledged the wet with a shrug, "It's a damp morning, but I've seen it a sight wetter'n this on the Old Oregon Trail and we didn't stop 'til we mired."³⁸

What turned out to be his last trip over the Old Oregon Trail was a smashing success from Meeker's prospective. The view from New York was not as sanguine. After picking up the expenses of Meeker's

Meeker in Tacoma, Washington, selling coins in November 1926. *Washington State Historical Society, 1957.19.261029*

jaunt, Howard Driggs noted years later that OTMA actually lost ten thousand dollars on the campaign.[39]

When the coin was finally issued, opinions varied on its appearance. Robert Bruce was disappointed as he felt "Indian is overlarge in proportion to wagon" and that it obscured the map of the route.[40] Bagley, however, thought the coin was beautiful, but complained about the similar illustration Bruce had designed for the OTMA letterhead because the oxen were so large you couldn't see Meeker in the wagon.[41] Bruce promised a revision in the near future. In response to Bagley's request for numbers 1843, 1852, and 1865, he replied that the first one hundred coins minted were in individual envelopes in vaults of the Equitable Trust Company but that the remainder were not numbered in any fashion. A plan for a White House ceremony where Ezra presented coin number one to President Coolidge, and Coolidge handed it back in exchange for coin number two, never materialized. Minnie Howard ended up with coin number two, which today is stored in the Idaho State Museum. The whereabouts of Meeker's coin number one is unknown. Bruce promised to obtain numbers 43, 52, and 65 for Bagley.[42]

Trouble in New York

The fledgling Oregon Trail Memorial Association was not yet one year old, and already trouble was brewing. On October 20, 1926, Robert Bruce sent a long letter to William Bonney describing tensions between the OTMA office and Meeker. Bruce told Bonney that Meeker was very fortunate in having men like Seymour, Barber, Rogers, and Davis backing him and that all were distressed by the barrage of multiple page letters that Meeker directed at the board members nearly daily, lambasting their decisions. Bruce wrote that Meeker misread a letter from Seymour regarding the process established for distribution of the coins. Subsequent letters and telegrams failed to "shake this out of his head," and that Meeker was bombarding the board with threats to return and "have it out" with the executive committee. "He would help us much more by staying in the Northwest, as while he is here it takes the whole office force to wait on him—and we can only accomplish anything after he has gone for the day," continued Bruce. "Personally, I am all fed up, and would have left the work long ago except for the feeling that something might suffer." He informed Bonney that he was sending a copy of the letter to Clarence Bagley. Bruce concluded by saying,

What this office needs to carry out its work in the very best and most efficient way is a freedom from constant nagging, and for Uncle Ezra to go along in his own best sphere of meeting and personally interesting the public. If by any means this thought could be put over to him in a friendly way, it would help us very much; but of course this letter is confidential between us, you and Mr. Bagley.[43]

Bruce continued to send Bonney and Bagley letters in this vein into November.[44] On November 23 Meeker left Seattle for New York to attend the annual OTMA general membership meeting. Meeker's arrival changed Bruce's tone. "Mr. Meeker's coming back here is already proving a wise move....The several days spent by Mr. Meeker here before this meeting, his talk with the bankers, and also with the individual members of the Committee, have, I am quite sure, convinced him that the procedure of this office has been for the best permanent good of the organization."[45] Bruce assured both Bonney and Bagley that the annual meeting was "friendly in character from beginning to end, and absolutely without any controversy."

The Clarence Bagley collection at the University of Washington holds a brief note from an L. Kitchel that may put Bruce's relationship with Meeker in a different perspective. It reads:

> Story regarding Bruce and Meeker has two sides. Bruce was discredited with his company by Meeker because of Bruce's frequent amours. Meeker adored his wife and had no toleration for such behavior. Bruce charged Meeker for editing his book, Kate Mulhall, more than Meeker thought was just, especially as Meeker had no funds and Bruce knew it. ALL Meeker's funds were used on Trail work for twenty years. He had no other needs.[46]

THE OREGON TRAIL MEMORIAL COIN

The friction between Meeker and the OTMA Board of Directors can be better understood with a description of the mechanics of selling memorial coins. Congress gave the U.S. Mint the authority to issue up to six million fifty-cent coins, with the OTMA as the only party authorized to sell them. The selling price would be one dollar per coin with the profit from the sales going to the OTMA for the purpose of marking the route of the Oregon Trail. The coins had to be purchased up front for fifty cents each and then shipped from the mint, the cost of shipping to be borne by the OTMA. The association's cash reserves were minimal, so they entered into an agreement with the Equitable Trust Company of 37 Wall Street, New York City.[47] Equitable advanced the cost of purchasing and shipping the coins—for a fee. It worked like this: The coin was sold for one dollar. The mint received fifty cents of that dollar and perhaps a nickel more to cover shipping costs. This money came from Equitable. The Equitable Trust Company took a percentage of the sale of each coin as payment for the loan—say fifteen cents per coin. That left the OTMA with a profit of around thirty cents per coin. The coins were shipped from the mint to Equitable upon the order of two OTMA officers. Equitable then delivered them to purchasers such as banks and collectors around the country. This too entailed a cost that was borne by the OTMA. From the perspective of the Board of Directors and the mint, this was the proper way to conduct the business. For the association it was the only way initially.

Meeker, of course, was out in the field actually selling the coins. He had learned from long experience selling his books and postcards that buyers wanted to see and touch the product they were being urged to purchase. He wanted a healthy supply of coins on hand and the ability to order more at a moment's notice. He asked for authority to purchase coins at will for fifty cents per coin, thus cutting The Equitable Trust Company out of the loop. This could not be done until the OTMA had received an influx of cash from the initial sales large enough to keep purchasing coins as Meeker sold them. By early 1927 Meeker felt this condition had been met.

Meeker lobbied non-stop for this authority, bombarding the individual board members with letter after letter. The board saw this as breaking their contract with Equitable and unanimously voted time and again to deny Meeker's request. Clarence Bagley in a January 20, 1927, letter to Robert Bruce summed up the problem in a nutshell. "How have you managed to keep him within bounds? Each passing year seems to increase his vitality and does add to his determination to have his own way."

Edmund Seymour spoke for the directors:

> About six weeks ago Mr. Meeker undertook to try to convince the Executive Committee of this Association that he should be allowed

to withdraw coins at will from banks at fifty cents each—which of course, would be actually below cost of minting, when considering the transportation charges. This we felt would be detrimental to the work of the Association as well as contrary to our arrangement with the Equitable Trust Co., and the Committee unanimously decided again and again not to accede to Mr. Meeker's wishes in this respect... [I]t would be a very strong point as well as a guarantee of good faith to the public, to be able to state that no one, not even the President or Ezra Meeker is able to secure a single coin at less than the established rate of $1.00.[48]

The directors calmed Meeker's fires for a time by setting up a $1,000 revolving fund that he could use to purchase coins when he wished.

Meeker spent the first months of 1927 lecturing and selling coins, primarily in the east. He claimed to have sold nine thousand coins in one week in April. But he complained constantly. In August he wrote Bonney, "I am bedeviled by [these] so-called Directors." Meanwhile, in Idaho, Minnie Howard was stirring up trouble. Under her guidance the Idaho OTMA chapter asked the directors for authority to purchase 200,000 coins at fifty cents per coin, with the profits directed entirely to the proposed Fort Hall monument. In a stunning display of hypocrisy Meeker objected, saying, "This proposition to set out 200,000 under the independent control of another combination would be the beginning of the end of this Association."[49] It was exactly the conditions he was proposing for himself.

By the end of the year Meeker was scheming to replace the New York OTMA board and move the headquarters to Seattle. He asked Bonney to accept a position on this new board, which would apparently consist of his daughter and grandson, and other loyalists. "I want you to accept as one of the directors and cooperate with me when I establish the office in Seattle along with Joe Templeton, Mrs. Osborne and Fitzgerald."[50] This scheme failed.

In the end 264,250 coins were delivered to the OTMA and a disappointing 61,244 went unsold. These were eventually returned to the mint to be melted. In 1939 Congress withdrew permission for the mint to release any further memorial coins, including the Oregon Trail fifty-cent piece. The profits for the OTMA were probably around $60,000–$70,000, nowhere near the three million dollars Meeker envisioned when the bill was passed in 1926.[51]

Fraud?

Some coin historians have questioned the integrity of the OTMA leaders in regards to the Oregon Trail Memorial coin. Coin historian Q. David Bowers wrote, "As far as I know, the financial benefits which provided the reason for issuing the half dollars, 'to rescue the various important points along the old trail from oblivion, to erect suitable monuments, memorial and otherwise,' etc., never came to pass, at least not from money provided by the sale of the coins."[52]

This judgment by Bowers, and others before him, implying fraud, became the accepted version in numismatic circles.[53] Some numismatics argue that this accusation led to Congress' decision to end the commemorative coin series in 1939.[54]

Gary Greenbaum's articles in the *The Numismatist* effectively dispute this condemnation and set the record straight.[55] It would appear Bowers never traveled the route of the trail, for surely it would be hard to miss the work done by the OTMA, financed in part by its memorial coin. Anyone who travels the route today will see well over one hundred OTMA markers at important sites along the route of the pioneers from Washington, through Oregon, Idaho, Wyoming, Nebraska, Kansas, and Missouri. They all feature a bronze plaque illustrated with a covered wagon traveling toward a setting sun, with the words "Oregon Trail Memorial" inscribed below—the same image that is on the memorial coin. The monuments both mark the route of the trail and memorialize the pioneers, which was exactly the OTMA's intention. Bert Webber, the author of *Ezra Meeker: Champion of the Oregon Trail*, wrote in his booklet *The Oregon Trail Memorial Half-Dollar (1926–1929)* about a missing OTMA monument funded in part by students of Yale University who purchased six hundred of the half-dollar coins in 1930. Webber traveled to Yale, Idaho, where the monument was supposedly placed, searching the ground, and checking various archives. Like Bowers, he asked what had happened to the money. His judgment, however, was less severe and left room for doubt: "No Oregon Trail Marker financed by money from Yale University students, faculty and friends, was ever installed at Yale, Idaho. But hard judgments, even in the face of all the 'no marker' evidence, should probably be withheld for there is yet the plausibility someone we don't know to contact, might eventually lead us to the marker."[56] On U.S. Highway 410 in the hamlet of Greenwater, Washington, there is an Oregon Trail marker consisting of a pyramid of cemented stones in which is embedded OTMA's distinctive bronze medallion. A second plaque names all the immigrants who came over Naches Pass in the first wagon train to traverse the route in 1853. A third plaque reads: "Marker erected in memory of pioneers who came over the Naches Branch of the Oregon Trail. Financed by students of Yale College, Pierce County Pioneer Society and other friends. Sponsored by Washington State Historical Society 1941." Monument found, but mystery not quite solved.

A May 18, 1931, letter to William Bonney advised that the directors of the OTMA had placed four markers in Washington State and that they were anxious to place fifteen more, including one at "the point where the Green Water River enters the White River."[57] Bonney responded, "We have one in our office, furnished by the Students of Yale College; it is proposed to place this one at Green Water, but the High-way is to be changed there, and the Commission will not issue a permit to us until the engineers have finished the change of line."[58] The monument, funded by the Yale students, stands on the site where Edward Jay Allen, builder of the Naches Pass wagon road, set up his work camp. If the marker was originally intended for installation in Yale, Idaho, how it came to be selected instead for a site in Washington remains a mystery, but clearly no fraud was committed.

CHAPTER 20

End of the Trail

As the final year of his busy life dawned Ezra Meeker's frustration with the Oregon Trail Memorial Association board of directors did not abate. On May 11 he wrote William Bonney again about his plan to put his daughter Carrie Osborne, his grandson Joe Templeton, and old friend Maurice Fitzgerald on the board and move the headquarters to Seattle. He again urged Bonney to run for a seat on this newly constituted board and wrote, "I can hold the next annual meeting in Seattle for the election of Directors but if I am not sustained by the people of Tacoma and Seattle I am going to walk out of this office and throw up my hands beaten."[1] On May 20 Bonney begged off, pleading lack of time to give the position the attention it deserved. In July Meeker sent a letter to the general membership stating, "I now realize that I made a grave mistake in incorporating the Oregon Trail Memorial Association in New York which has resulted in two years of contention that has finally told upon my vitality so that I now have no other choice than to go to my home-town Seattle where I will be surrounded by my many friends and numerous descendants." He went on to state that the OTMA by-laws allowed him to call a special meeting of the members upon the request of 25 percent of the membership. He declared his intention to call such a meeting in Seattle on the second Monday of December. Enclosed was a form by which the members could make such a request. He concluded by stating, "Particulars of the objects of the special meeting will be set forth in the formal notice which will follow."[2]

No doubt to the relief of the board, Meeker got involved in another project that ate up his time. Henry Ford donated the chassis of one of his trucks valued at $900 to the association, and funds were provided to Meeker to turn it into another motorized old-time prairie schooner. Meeker's intention was to take it from New York and travel over the Oregon Trail to Seattle, arriving in late November.[3] Howard

Driggs wrote that Meeker virtually lived in the upper Manhattan shop where he supervised its construction. Meeker's correspondence confirms Driggs' recollection. He solicited a bid to build the wagon box and bonnet to place on the chassis from a local company and found the price exorbitant. He turned to Bonney and asked him to contact Joseph R. Turner, president of the steel company that built his 1923 wagon, to ascertain what they would charge to build a wagon box and ship it east. Meeker eventually found a local company to build the wagon body and asked Bonney to supply the measurements of the 1906 wagon displayed in the Tacoma museum glass case.[4]

The "Oxmobile," as it was called, was ready by the end of July. On August 5 Meeker and driver left New York City for a short tour of New England, a section of the country that he had neglected until now. This corner of the country had produced persons of note in western history such as Robert Gray, discoverer of the Columbia River; Nathaniel Wyeth, founder of Fort Hall; Eliza Spalding, among the first missionaries to cross the plains by wagon; and the celebrated "Mercer girls," marriageable women who came by ship to the Northwest in the 1860s to start new lives. While traveling through Sudbury,

Meeker with his 1928 Oxmobile in Detroit just before his final illness. *Seattle Museum of History and Industry 1986.5G.1920*

Massachusetts, Meeker met Henry Ford and his wife, who were staying at the historic Wayside Inn there. After an exchange of courtesies, Mr. Ford remarked, "You must bring your Oxmobile out to my plant at Dearborn and let my men fit it up with Lincoln shock-absorbers and other conveniences. I'd like to make it more comfortable for a man of your years."[5] Meeker accepted the offer.

The *New York Times* said this trip was "to be a preliminary to Mr. Meeker's sixth transcontinental trip over the Oregon Trail.[6] Both tours are being made in the interest of the OTMA... Mr. Meeker expects to return to New York Aug 27, and to start shortly after on his transcontinental trip."[7] Ralph Steen, a twenty-five-year-old University of Kansas student, was the driver and Meeker's aide. The plan was to camp out, sleep, and cook in the Oxmobile, lecture at various stops, and sell the coins wherever possible. Meeker left the city with a Hoover button pinned to his lapel and offered duplicates to those who saw him off at the Ford headquarters building. Driggs stated that even though the trip received good press and notice, it was not a financial success.

Rose Jay Schwartz, secretary of the OTMA, took a train to Dearborn to make arrangements for Meeker's reception. Unfortunately, Meeker arrived gravely ill and Ford immediately sent him to the Henry Ford Hospital where he stayed from September 13 to October 14 wavering between life and death. Reaching deep into his amazing vitality Meeker walked out of the hospital in mid-October. The hospital bill for Case No. 116160 was $837.22, paid by Henry Ford.[8] Upon his discharge Meeker went straight to the Oxmobile where a series of photographs were taken of him and his modern prairie schooner.

Guthrie Barber, the treasurer of the OTMA, went to Michigan to assist Ms. Schwartz in overseeing Meeker's care and the two of them talked Meeker out of his aim to continue his journey. Barber purchased a railroad ticket and secured a stateroom in a Pullman car for the ailing pioneer and Meeker made his final journey home.[9]

Ezra Meeker arrived in Seattle on October 31 after a nearly two-year absence. He brushed off his recent severe illness saying he had just walked a third of a mile for exercise—this despite thirty-seven days in the hospital where he took more medicines than he had taken in his entire life. Instead he talked to the reporters who met him about "every subject from aviation to women's short skirts—and, by the way, he likes 'em short—and then settled down to the big subject that is harassing him right now—will he be able to vote at the election Tuesday?"¹⁰ Meeker was the only surviving voter of those who cast a ballot in the first Washington territorial election in 1854. He had not missed a presidential election since Washington became a state, but he had been away so long that his voter registration had lapsed. His daughter Carrie Osborne said her father was greatly concerned about the possible loss of his vote and she was going to confer with professor Edmund Meany and Clarence Bagley to see if something could be done. Meeker then posed for a picture eating a Washington-grown apple in his room at the Frye Hotel¹¹ as it was Apple Week.¹²

The respite was short-lived. By the end of November he was again near death and was being kept alive by intravenous injections of a saline solution and glucose. Meeker was unable to speak except at rare intervals and then only in a feeble whisper. His last words whispered to his daughter Ella Templeton were, "I cannot go yet. My work is not finished."¹³ Ezra Manning Meeker died at four o'clock in the morning on Monday, December 3, 1928, just twenty-five days short of his ninety-eighth birthday.

Meeker's 1923 wagon picking up flowers to carry to the funeral home where his body lay in state. *Author's collection*

Meeker's death was announced in newspapers from coast to coast. That day two local pilots took off from nearby Boeing Field, flew over the Frye Hotel, dipped their wings in salute, and released a rain of rhododendrons blossoms on the hotel and crowds in the surrounding street.[14] On Tuesday morning Meeker was laid in state at the Hamilton Mortuary. In a fitting tribute to the old pioneer, the prairie schooner built for his 1923 motion picture effort was brought to Seattle and filled with flowers.[15] Pulled by two oxen and led by two mounted policemen, the wagon slowly made its way up Third Avenue to Olive Street, where it began the long uphill climb to the mortuary where it delivered its bouquets.

Through the afternoon and evening a steady stream of visitors came to pay their respects, among whom were scores of school children calling with flowers. The funeral was held Wednesday morning at the Westminster Presbyterian Church in Seattle with Ezra Meeker's grandson, Reverend Harry S. Templeton, and Reverend Lincoln Smith, an old friend, presiding. The honorary pallbearers were Clarence B. Bagley, George H. Himes, Maurice Fitzgerald, William Bonney, and Charles Ross, all dear close friends.[16] In the afternoon he was taken from the church by hearse to Puyallup, where the cortege stopped briefly in front of his Pioneer Park statue and the site of his first Puyallup home. Then it was on to Woodbine Cemetery, where he was laid to rest beside his beloved wife Eliza Jane.

All of his former Puyallup friends and business associates were there, as city businesses closed during the hours of the funeral. In 1939 the Oregon Trail Memorial Association placed a monument on the graves of the pioneer couple. It reads in part "They Came This Way to Win and Hold the West." Above their names is the bronze OTMA medallion featuring Eliza Meeker and her firstborn sitting in the wagon, and Ezra leading it west toward the setting sun.

Meeker's grave marker in Woodbine Cemetery, Puyallup, with OTMA bronze tablet. *P. Ziobron photograph*

Afterword

On April 16, 2018, I received the following email from Andy Anderson, then president of the Puyallup Historical Society at the Meeker Mansion. Andy is a longtime Meeker scholar and he posed an interesting Question:

> I keep wondering where the entrepreneurial spark came from, and where the beginnings of his business acumen came from. Certainly being a printer's devil at a very young age helped him understand the business of printing, but what taught him the value of advertising? In 1858 he got wind of the Fraser River gold rush out of Bellingham. What possessed him to amass a herd of cattle, raft them to "Whatcom" as it was then known, and land them on a piece of beach to run a dairy? Printing his own book in some 5000 copies, taking the train back to New York and seeking out Jay Cook; becoming Mr. Cook's live specimen for some months and then returning west to take up hop farming again was a step few might have taken…his whole progress through business after business, each building on the other is spellbinding. But still, where was the spark?

After more than two decades of researching, writing, and lecturing about Ezra Meeker I believe I have unraveled many of the characteristics of his personality that drove him—a willingness to take risks and embrace the new, a boundless capacity for work, an unrelenting competitive drive, a strict moral compass, and an unerring vision of the future. Personal comfort and the trappings of wealth were not high on the list of things that were important to him. These traits, coupled with remarkable good health and longevity, propelled him to heights that were no doubt unimaginable to the young man who came west over the Oregon Trail in 1852. His pugnaciousness and tendency toward self-righteousness made him difficult to work with at times, but this did not stop him or even slow him down. But where did the spark come from?

A look at the Meeker ancestors perhaps might help in part to answer the question. In them we see a willingness over generations to take great risks, and a tendency toward pugnaciousness. We start with

William Meeker, Ezra's sixth great-grandfather (1620–90), the first of the clan to migrate to America, arriving in Massachusetts around 1635 as an indentured servant. Four years later he was in New Haven, Connecticut, helping to build that colony. In 1646 he married the well-connected and wealthy Sarah Preston whose father was a signer of the January 24, 1639, "Fundamental Orders of Connecticut." By 1665 the family was in New Jersey, and over the next decade William Meeker became one of the leaders in the turmoil that eventually drove the British appointed governor James Carteret out of the colony to seek refuge in England. Ezra's third great-grandfather, Timothy Meeker (1708–98) fought in the American Revolution, enlisting at age seventy and bringing into his company his nine sons, two sons-in-law, and a grandson. His grandfather, Manning Meeker (1774–1816) started the family's westward movement, settling in New York. Ezra's father, Jacob (1804–69), continued the movement west to Ohio and eventually Indiana.

Clearly Ezra came from a long line of risk takers. Confronting the existing political order and pushing for change was something his family had been doing for generations. Striking out boldly on a new enterprise held no terror for him. In his various books Ezra also described his parents' capacity to work long, hard hours. This too was a trait that he inherited.

The "spark" Mr. Anderson reflected upon apparently began at birth and developed through Meeker's childhood. As an adult he was merely fine-tuning. Meeker wrote that he "was restless from the beginning—born so." At age nine Ezra walked two hundred miles from his home near Cincinnati, Ohio, to Attica, Indiana, behind the family's covered wagon. At age eleven, when the family moved to Indianapolis, Ezra began working at odd jobs. Over the next three years he saved thirty-seven of the fifty dollars he needed for a hoped-for purchase of forty acres of farmland in Hendrix County, hinting at that single-minded drive that propelled his future actions. His moral compass was also formed early when one of those jobs led him, at age thirteen, to the *Indianapolis Journal*, a local Free Soil newspaper. There he was introduced to the raging debate over slavery and the lynching of an African-American on an Indianapolis street. That year (1844) he attended a Whig convention and was bold enough to sing campaign songs to the attendees and sell them copies of the lyrics. At age fourteen his father put him in charge of the family farm while he worked in a nearby mill. It seems that Meeker was born with the spark embedded in his genes and by age fourteen it was fully engaged.

One trait that came to full fruition in Ezra was his genius for self-promotion. He became a master of the skill, at times bending the truth if it served his purpose. An example of this is his version of the origins of his 1852 trek over the Oregon Trail. In all his writings Meeker gives the reader the impression that the impetus for the journey west was his, when in fact his brother Oliver and certain members of the Ballard family who were living in Indianapolis initiated the plan. Oliver stopped in Iowa on his way west and basically scooped up his younger brother. Meeker also failed to mention that the Ballard party traveled a good way along the trail with them and that David Ballard, one of his trail mates, eventually became a future governor of Idaho Territory.[1] Disclosing this information would, of course, remove a major portion of the spotlight from himself. When writing of the 1852 trip west in *Pioneer Reminiscences*, he used fellow-traveler Eliza McAuley's diary as a reference. When she pointed out an error in his recounting the length of time he spent at Big Hill in Idaho, his response was telling. "Maybe the cold facts will spoil a nice little story I have written. I am not sure but will let it remain but the incident shows how much we are at fault in memory of things occurring half a century ago." He let it remain.[2] Similarly he wrote in *The Busy Life of Eighty-Five Years of Ezra Meeker* of floating through Miles Canyon and the Whitehorse Rapids on his way to the Yukon in 1898. The "ducking" he described when going through the rapids, leaving the reader to believe he was soaked by the spray from the turbulent waters, was explained in a contemporary letter to his wife as coming from a rather ordinary rainsquall instead.[3]

I have tried throughout the various volumes of this biography to define the characteristics that made up Ezra Meeker, both the good and bad, but I still occasionally get questions such as, "I know he did all these amazing things, but what about his dark side, his unpleasant qualities?" Yes, they were there, as in all of us, but they were not the dominant part of his personality. He was not cut in the mold of the robber barons of the Gilded Age or some of the more notorious members of present-day corporate culture, willing to toss all decency aside in pursuit of an end. He was at his core a moral man. In his life as a businessman he was a fierce competitor, at times too fierce. In his

Oregon Trail persona he brought this single-mindedness and competitiveness to his crusade. At times he alienated people. His various quests took a toll on his family members, friends, and acquaintances. But it was never personal. He would battle tooth and tong with an opponent or a family member over an issue, but he was willing to sit down to dinner with them when the contest was over.

His conflicts with Edmund Meany illustrate this well. He clashed with Meany over control of the Washington State exhibit at the Chicago World's Fair in 1893, going so far as to accuse Meany of fraud. Meany won the battle and had Meeker removed from his role as CEO of the exhibit. In 1905 the two men again clashed, this time in the pages of the *Seattle Times*, over Meeker's portrayal of Governor Isaac Stevens in his book *Pioneer Reminiscences*. But in 1906 when Meeker needed someone to handle the money he was raising to erect monuments along the Cowlitz Trail in Washington, he turned to Meany to manage the funds. Another example is Meeker's behavior in the aftermath of the collapse of Washington's hop industry in the 1890s. Meeker was left holding debts of some $100,000 owed to him by his fellow hop growers. Despite facing his own bankruptcy, he made no effort to collect any of that debt, as he knew his neighbors were in equally desperate straits. Another type of man might have gone after whatever he could get to save himself.

Perhaps his greatest personal failing was the time he spent away from his wife and the burden that placed on her. Their fifty-eight-year marriage was, in part, defined by long separations. That he loved his wife was obvious from his writings. It is clear from the historical record that Eliza Jane was a partner in his various enterprises. It is equally clear that she was at times lonely during his absences. When she slipped into dementia Meeker was consumed with guilt about his absences. He rationalized that he was earning money needed for her care, and there were limited opportunities to do so at his advanced age. In truth, he loved what he was doing, and nothing short of his own death would stop him. One could criticize him for these choices, but we need to remember there was no social safety net in those times. Meeker was operating under rules much different from those of today.

Meeker's ultimate quest was not simply to line the trail with monuments and to save what remnants of it could be saved. Meeker explained the purpose of the monuments: "By putting up monuments along this historic trail, I knew that the children of generations yet unborn would ask their elders, 'What mean ye by these stones?' and in answer to their questions, would hear the story of the crossing of the plains and the settlement of the West, and hearing it, would value more highly the heritage won for them by their fathers." Howard Driggs put it more simply. "His goal was the preservation of our historical heritage." Today, in addition to monuments, multiple interpretive centers dot the route. The story of the Oregon Trail is well told.

How did Native communities fit into this celebration of the pioneers? Meeker defined his relationship with Native Americans beginning with his adoption of a Native American orphan and the Washington Territory conflicts of the 1850s.[4] He argued in print and in person that he felt the tribes had been wronged when put on reservations too small and too poor to provide a living. He argued that Chief Leschi had been judicially murdered and never wavered in that belief. His working relationship with Native Americans during his hop growing years was amicable. He hired thousands as pickers and field hands. But he was a man of his times. The story he wanted told was that of the pioneers. He gave years of lectures and penned page after page on that topic. From 1909 to 1925 the only extensive interactions he had with Native Americans were at the Alaska-Yukon-Pacific Exposition in 1909 when he hired a Native group to perform daily at his exhibit, and, in 1925, at the 101 Ranch Wild West Show where he was attacked nightly by arrow-shooting Indians as he made his escape with his covered wagon into the safety of the fort.

In his decades of lecturing he followed a pattern. His pamphlet, *Story of the Lost Trail to Oregon*, of which he sold or gave away in the thousands, and his speech to the YMCA printed in the Springfield, *Illinois State Register* on November 6, 1910, are prime examples. He began with,

> No more fascinating search can ever engage the mind of man than to follow the actual track of the fathers that opened the way to conquer the land and wrest it from the native race and erect the standard of civilization in the wilderness and with the standard, a barrier to an encroaching nation [Great Britain] whose grasp was tightening upon an empire—The Oregon Country.... The Indians and the buffalo possess the land; we cannot see that manifest destiny has set the seal upon the destruction of the one [buffalo] though we may dimly realize, the power of the other is destined to be broken in the march of civilization.

That pretty much summed up his views on the inevitability of Manifest Destiny, and he did not dwell on the topic for long in any speech or writing. The meat of his lectures was always the trials and suffering of the pioneers on that six-month journey across the plains to Oregon—the human drama of the exodus. This is what he wanted people to remember. The dust, the cholera, the graves, and the teamwork it took to make the journey. Meeker's Oregon Trail quest came during a time when Native Americans were more often than not viewed in stereotype. The Wild West shows full of such stereotypes were at their zenith. Early motion pictures incorrectly portrayed the trail experience as full of endless conflict with Native Americans. Today, we know better, and many of the interpretive centers along the route tell a more complete story. Under the leadership of Howard Driggs, the Oregon Trail Memorial Association took up Meeker's work after his passing, placing monuments along the trail and sponsoring local and national trail celebrations until December 29, 1939. Its members then formed a new organization titled the American Pioneer Trails Association (APTA) that lasted until 1954. This new organization stayed active during World War II, promoting the centennial celebration of the "Great Migration of 1843" at localities around the nation. Part of APTA's legacy was a survey done by the Oregon Highway Department to locate and document the remnants of the trail in Oregon in preparation for its statehood centennial celebration in 1959.

In 1968 Congress passed the National Trails System Act, and in 1978 National Historic Trails designations were added. Today there are eleven National Historic Trails. The system generally consists of remnant sites and trail segments and promotes their preservation, interpretation, and appreciation. While administered by federal agencies, land ownership may be public or private. On August 12, 1982, the Oregon-California Trails Association (OCTA) was created and considers itself the offspring of Meeker's OTMA.

Meeker's great legacy to America, the saving of the Oregon Trail, is ongoing. Today the trail is lined from end to end with monuments, many dating from his time. Some are in rural areas, some in highly developed towns and cities and along busy streets and highways. Recognition as a National Historic Trail didn't automatically confer protection. Precious few sections of the trail remain, and of those even fewer are as pristine as they were in 1852. There are, and have always been, continuing pressures to develop the land it transects, both on public and private property. That pressure will only increase into the future. OCTA has adopted Meeker's mission and continues his work to save the trail where possible, and to educate the public as to its role in American history. Other groups whose goals are similar have formed since. Trails West Inc. formed in 1970, focusing primarily on trails to California, some of which are shared with the Oregon Trail, and with a mission similar to OCTA's. The Applegate Trails Association is active in southern Oregon. And OCTA has local chapters all along the route of the Oregon Trail, as well as along the various trails to California. All of these are a product of Ezra Meeker's gift to America.

Notes

INTRODUCTION

1. This was the professional title that Driggs, a professor of English at the University of Utah (1907–23) and New York University (1927–42), and prolific author of Western-themed fiction and nonfiction, preferred to use. Ezra Meeker and Howard Driggs, *Ox-Team Days on the Oregon Trail* (Yonkers-on-Hudson, NY: World Book Company, 1925).

2. Howard Driggs, *Covered-Wagon Centennial and Ox-Team Days* (Yonkers-on-Hudson, NY: World Book Company, 1931), 317.

CHAPTER 1: TELLING HISTORY HIS WAY

1. "Rebuilds a Fortune for a Golden Wedding," *San Francisco Call*, May 3, 1901, 6.

2. Ezra Meeker to Eliza Jane Meeker, September 11, 1902, Meeker Papers, Box 4, Folder 2A.

3. Ezra Meeker to Ella Templeton, April 30, 1903, Meeker Papers, Box 6, Vol. 1.

4. An inflation calculator (www.in2013dollars.com/1909-dollars-in-2016?amount=0), shows that these were not trivial amounts of money. The $600 debt in 1903 would be $15,645 in today's currency.

5. Ezra Meeker to Molly Male, February 13, 1903, Meeker Papers, Box 6, Vol. 2.

6. Ezra Meeker to Edward N. Fuller, December 26, 1903 Meeker Papers, Box 6, Vol. 1.

7. Ezra Meeker to Ella Templeton, April 30, 1903, Meeker Papers, Box 6, Vol. 2.

8. Ezra Meeker to Robert Wilson, October 13, 1903, Meeker Papers, Box 6, Vol. 1.

9. The story of the portrait is told in Dennis M. Larsen, "A Fatal Intersection," *Tumwater Historical Association Newsletter*, Vol. 35, Issue 3 (Winter 2016), and in a letter, Ezra Meeker to Messrs. Benson & Morris Co., October 18, 1903, Meeker Papers, Box 6, Vol. 1.

10. Ezra Meeker to Eliza Jane Meeker, October 29, 1903, Meeker Papers, Box 6, Vol. 1.

11. Ezra Meeker to Henry Hewitt, November 24, 1903, Meeker Papers, Box 6, Vol. 1.

12. Ezra Meeker to William Billings, December 21, 1903, Meeker Papers, Box 6, Vol. 1.

13. Ezra Meeker to Robert Wilson, January 28, 1904, Meeker Papers, Box 6, Vol. 1.

14. Ezra Meeker to Arthur E. Grafton, January 1, 1905, Meeker Papers, Box 6, Vol. 1.

15. See Pulitzer Prize-winner Richard Klugar's *The Bitter Waters of Medicine Creek*

(New York: Vintage Books, 2011), 81–82 for the various contradictory accounts. For a different account see Kent Richards, *Isaac I. Stevens: Young Man in a Hurry* (Pullman: Washington State University Press, 1993), 197–202.

16. Ezra Meeker to A. G. Foster, November 19, 1904, Meeker Papers, Box 6, Vol. 1; Ezra Meeker to James H. Spalding, November 20, 1904, Meeker Papers, Box 6, Vol. 1; Ezra Meeker to A. C. Towner, December 23, 1904, Meeker Papers, Box 6, Vol. 1.

17. Ezra Meeker to Eva Gear, December 28, 1904, Meeker Papers, Box 6, Vol. 1.

18. Ezra Meeker to Katherine Graham, January 23, 1905, Meeker Papers, Box 6, Vol. 1.

19. Ezra Meeker to Roderick McDonald, December 12, 1904, Meeker Papers, Box 6, Vol. 1.

20. Louise Brown Ackerson was a San Francisco schoolteacher and John W. Ackerson's second wife. She was active in Tacoma community affairs and argued on December 16, 1890, and July 24, 1903, in letters to the *Seattle Post-Intelligencer* that her husband was the first to confer the name Tacoma on the fledgling city. Herbert Hunt, *Tacoma, Its History and Its Builders: A Half Century of Activity, Volume 1* (Chicago: S. J. Clarke Publishing Company, 1916), 128 and 137.

21. Ezra Meeker to Louise Ackerson, December 30, 1904, Box 6, Vol. 1.

22. Ezra Meeker to Louise Ackerson, December 31, 1904, Box 6, Vol. 1.

23. Ezra Meeker to Louise Ackerson, January 4, 1905, Meeker Papers, Box 6, Vol. 1.

24. The Metropolitan Press was a Seattle based printing company owned by Henry C. Pigott. The company published the *Oregon Historical Society Quarterly* and legal papers for the Washington State government; Ezra Meeker to Louise Ackerson, January 6, 1905, Meeker Papers, Box 6, Vol. 1.

25. Ezra Meeker to Louise Ackerson, February 22, 1905, Meeker Papers, Box 6, Vol. 1.

26. Ezra Meeker's *Pioneer Reminiscences of Puget Sound: The Tragedy of Leschi* (Seattle: Loman & Hanford, 1905) may be read online at Google Books.

Chapter 2: A War of Words, a Test of Wills

1. "Meeker Not Like Cicero, Says Meany," *Seattle Times*, April 15, 1905, 4.

2. See Dennis M. Larsen, *Hop King: Ezra Meeker's Boom Years* (Pullman: Washington State University Press, 2016), 137–39, for the story of that clash.

3. "Offers to Prove Charges," *Seattle Times*, April 26, 1905, 7.

4. Gibbs' letter originally appeared in June 5, 1857, issue of the *Washington Republican*. Meeker reprinted it in *Pioneer Reminiscences of Puget Sound: The Tragedy of Leschi*, 258–59.

5. "Charge of Malice Too Mild," *Seattle Times*, May 1, 1905, 3.

6. See chapter 3, Origins of the Monument Expedition.

7. Meeker also started his 1906 Monument Expedition with this Puyallup couple. Mrs. Gobel was five months pregnant in May 1905.

8. The horse team belonged to George Cline who accompanied Meeker to Skagway, Alaska, in 1898, and helped him ship his vegetables up Chilkoot Pass. Cline ran a livery stable in Puyallup.

9. The camp was near the corner of 24th and Thurston. June 12, 1905, Journal entry, Meeker Papers, Box 8, Folder 5E.

10. June 7, 1905, Journal entry, Meeker Papers, Box 8, Folder 5E.

11. June 10, 1905, Journal entry, Meeker Papers, Box 8, Folder 5E.

12. Ibid.

13. June 12, 1905, Journal entry, Meeker Papers, Box 8, Folder 5E.

14. July 25, 1905, Journal entry, Meeker Papers, Box 8, Folder 5E. Dillon S. Farrell was a dealer in General Merchandise, Paints and Oil in Toledo, Washington. Later he became the Toledo postmaster and mayor. Farrell arranged to have the oxen pastured on a farm owned by a Mr. Baxter. D. S. Farrell to Ezra Meeker, September 4, 1905, Meeker Papers, Box 7, Folder 3.

15. Ezra Meeker to Katherine Graham, September 29, 1905, Meeker Papers, Box 7, Folder 3.

16. A. C. Going to Ezra Meeker, July 12, 1905, Meeker Papers Box 7, Folder 3. Going ran a stove and range business in Portland.

17. Ezra Meeker to Carrie Osborne, July 20, 1905, Meeker Papers, Box 4, Folder 2B.

18. August 31, 1905, Journal entry, Meeker Papers, Box 8, Folder 5E.

19. September 1, 1905, Journal entry, Meeker Papers, Box 8, Folder 5E.

20. October 14, 1905, Journal entry, Meeker Papers, Box 8, Folder 5E.

21. September 4, 1905, Journal entry, Meeker Papers, Box 8, Folder 5E.

Chapter 3: The Old Oregon Trail Monument Expedition

1. Mr. and Mrs. Eben Ives to Ezra Meeker, June 27, 1905, Meeker Papers, Box 7, Folder 3.

2. The Gilliams Press to Ezra Meeker, June 29, 1905, Meeker Papers, Box 7, Folder 3.

3. Ezra Meeker to Jay Amos Barrett, August 15, 1905, Meeker Papers, Box 7, Folder 3.

4. Meeker inquired of the company that put up brass markers for the Sons of the American Revolution, but eventually came to realize that stone markers were more practical. William A. Hardy & Sons, Brass Founders, Fitchburg, M A, to Ezra Meeker, November 20, 1905, Meeker Papers, Box 5, Folder 20.

5. Meeker never explained where he obtained these slides.

6. The Cowlitz Trail was a branch of the Oregon Trail that led from Portland, Oregon, to Puget Sound.

7. See Sidebar: Building a Wagon, this chapter, for the story of the hub.

8. Ezra Meeker to R. S. Shakleford, January 9, 1906, Meeker Papers, Box 7, Folder 4A.

9. Ezra Meeker to George Himes, January 19, 1906, Meeker Papers, Box 7, Folder 4A.

10. Ibid.

11. Ezra Meeker to Lee Moorhouse, January 6, 1906, Meeker Papers, Box 7, Folder 4A.

12. Ezra Meeker to President Theodore Roosevelt, December 25, 1905, Box 6, Vol. 1.

13. The author misidentified Cora Mardon as Clara Mardon in *The Missing Chapters: The Untold Story of Ezra Meeker's Old Oregon Trail Monument Expedition* (Puyallup: Ezra Meeker Historical Society, 2006).

14. Ezra Meeker to Carrie (Meeker often called her Caddie) Osborne, May 14, 1908, Meeker Papers, Box 7, Folder 8C.

15. "Farm and Home," *Seattle Post-Intelligencer*, Sunday, October 16, 1887.

16. Ezra Meeker to Ella Templeton, October 13, 1903, Meeker Papers, Box 4, Folder 2.

17. Ezra Meeker to Carrie Osborne, November 26, 1906, Meeker Papers, Box 4, Folder 2B.

18. Ezra Meeker to Carrie Osborne, February 17, 1907, Meeker Papers, Box 4, Folder 2B.

19. The wagon Meeker used at the 1905 Lewis and Clark Exposition appears to have been a modified farm wagon that was not used again.

20. Ezra Meeker to George Himes, January 19, 1906 Meeker Papers, Box 4, Folder 12A.

21. October 20, 1905, Journal Memorandum, Meeker Papers, Box 8, Folder 5E.

22. "Oct. 6, 1895, Independence, Oregon. [Lot Number] 2270 C Giesy 80 Bales" and "Oct. 14th, 1895 Aurora, Oregon [Lot Number] 2466 Dr. WW Giesy 40 Bales," Meeker Papers, Box 2.

23. Email communication from Nancy Jackson, Washington State Historical Society Research Center, May 2, 2017.

24. Ezra Meeker to the West Coast Wagon Company, November 20, 1905, Meeker Papers, Box 7, Folder 3.

25. Ezra Meeker to William Bonney, December 20, 1922, Meeker Papers, Box 4, Folder 7C.

26. Ezra Meeker to T. Barley, January 15, 1906, Meeker Papers, Box 7, Folder 4A.

CHAPTER 4: DISASTER AT THE ALASKA-YUKON-PACIFIC EXPOSITION

1. See Larsen, *Hop King*, 199–202.

2. Ezra Meeker to Carrie Osborne, December 25, 1907, Meeker Papers, Box 4, Folder 5.

3. Ezra Meeker to Ira A. Nadeau, January 4, 1908, Meeker Papers, Box 4, Folder 13.

4. Ira A. Nadeau to Ezra Meeker, January 10, 1908, Meeker Papers, Box 4, Folder 13.

5. Ezra Meeker to Ira A. Nadeau, January 17, 1908, Meeker Papers, Box 4, Folder 13.

Notes for pages 39–51 229

6. A. W. Lewis to Ezra Meeker, February 24, 1908, Meeker Papers, Box 4, Folder 13.

7. Henry E. Dosch, Director of Exhibits and Privileges, to Ezra Meeker, August 11, 1908, Meeker Papers, Box 4, Folder 13.

8. Ezra Meeker to Ira A. Nadeau, November 18, 1908, Meeker Papers, Box 4, Folder 13.

9. Ezra Meeker to Ira A. Nadeau, November 25, 1908, Meeker Papers, Box 4, Folder 13.

10. Ezra Meeker to George Himes, January 2, 1909, Meeker Papers, Box 6, Vol. 3.

11. Ezra Meeker to John McGraw, January 5, 1909, Meeker Papers, Box 6, Vol. 3.

12. Ezra Meeker to George Himes, May 3, 1909, Meeker Papers, Box 6, Vol. 4.

13. Ezra Meeker to Marion Meeker, January 6, 1909, Meeker Papers, Box 6, Vol. 3.

14. Ezra Meeker to W. H. Paulhamus, March 28, 1909, Meeker Papers, Box 6, Vol. 4.

15. Ezra Meeker to Jenny Lester Hill, September 4, 1909, Meeker Papers, Box 6, Vol. 5.

16. Ezra Meeker to Mayor Joseph Simon, August 26, 1909, Meeker Papers, Box 7, File 9.

17. Ezra Meeker to William Mardon, September 2, 1909, Meeker Papers, Box 6, Vol. 5.

18. Ezra Meeker to William Mardon, September 10, 1909, Meeker papers, Box 6, Vol. 5.

19. Ezra Meeker to Jacob Price, September 22, 1909, Meeker Papers, Box 6, Vol. 5.

20. Ezra Meeker to Carrie Osborne, September 30, 1909, Meeker Papers, Box 7, File 9.

21. Ezra Meeker to Carrie Osborne, October 6, 1909, Meeker Papers, Box 7, File 9.

22. Ezra Meeker to Joe Kohler, October 7, 1909, Meeker Papers, Box 6, Vol. 5.

23. W. D. Simonds was pastor of the Oakland First Unitarian Church 1907–1918. His name is spelled Symonds in the correspondence. Ezra Meeker to Carrie Osborne, October 9, 1909, Meeker Papers, Box 7, Folder 9.

24. Today it is called Woodbine Cemetery.

25. Katherine Graham lived in the Puyallup mansion and was Eliza Jane's caretaker until Carrie Osborne took her mother to Seattle in 1907. Ezra Meeker to Katherine Graham, December 6, 1909, Meeker Papers, Box 5, Folder 6.

26. Ezra Meeker to Carrie Osborne, October 16, 1909, Meeker Papers, Box 6, Vol. 5.

27. Fanny Meeker to Ezra Meeker, October 13, 1909, Meeker Papers, Box 4, Folder 2D.

Chapter 5: Retreat to California

1. Ezra Meeker to Carrie Osborne, October 22, 1909, Meeker Papers, Box 7, Folder 9.

2. Ezra Meeker to Joe Templeton, October 31, 1909, Meeker Papers, Box 6, Vol. 6.

3. Ezra Meeker to Joe Templeton, November 8, 1909, Meeker Papers, Box 6, Vol. 6.

4. Ezra Meeker to Carrie Osborne, November 10, 1909, Meeker Papers, Box 6, Vol. 6.

5. Ezra Meeker to Clara Meeker, November 20, 1909, Meeker Papers, Box 6, Vol. 6.

6. Ezra Meeker to Carrie Osborne, November 26, 1909, Meeker Papers, Box 6, Vol. 6.

7. Ezra Meeker to Carrie Osborne, November 30, 1909, Meeker Papers, Box 6, Vol. 6.

8. Held at Dominguez Field in present day Carson, California, this was the first major air show in the United States. Attendance was listed as 254,000 over 11 days. Over 40 aviators competed for $75,000 in prize money. The Wright Brothers were there—not to fly, but to sue over a patent dispute. Meeker and Orville Wright would meet again in 1924 at the Dayton Ohio Air Show.

9. Ezra Meeker to Carrie Osborne, December 3, 1909, Meeker Papers, Box 6, Vol. 6.

10. Ezra Meeker to Katherine Graham, December 6, 1909, Meeker Papers, Box 6, Vol. 6.

11. Ezra Meeker to Carrie Osborne, November 26, 1909, Meeker Papers, Box 6, Vol. 6.

12. Ezra Meeker to Katherine Graham, December 6, 1909, Meeker Papers, Box 6, Vol. 6.

13. Ezra Meeker to Carrie Osborne, November 26, 1909, Meeker Papers, Box 6, Vol. 6.

14. Ezra Meeker to Joe Templeton, December 23, 1909, Meeker Papers, Box 6, Vol. 7.

15. Ezra Meeker to Carrie Osborne, December 26, 1909, Meeker Papers, Box 6, Vol. 7.

16. Ezra Meeker to Clarence Bagley, November 29, 1909, Meeker Papers, Box 6, Vol. 6.

17. Ezra Meeker to Carrie Osborne, December 29, 1909, Meeker Papers, Box 6, Vol. 7.

18. Ezra Meeker to Carrie Osborne, January 7, 1910, Meeker Papers, Box 6, Vol. 7.

19. Ezra Meeker to Carrie Osborne, January 13, 1910, Meeker Papers, Box 7, File 9.

20. Ezra Meeker to Joe Templeton, January 23, 1910, Meeker Papers, Box 6., Vol. 7.

21. Members of the Ballard family traveled with Meeker over the Oregon Trail in 1852. It is likely that Ling Ballard was related. Larsen, *Hop King*, 127 and 225.

22. Ezra Meeker to Carrie Osborne, January 24, 1910, Meeker Papers, Box 6, Vol. 7.

23. Ezra Meeker to Carrie Osborne, January 28, 1910, Meeker Papers, Box 6, Vol. 8.

24. Ezra Meeker to Carrie Osborne, February 7, 1910, Meeker Papers, Box 6, Vol. 8.

CHAPTER 6: THE SECOND OLD OREGON TRAIL EXPEDITION, 1910

1. The map may be found in Box 16 of the Meeker Papers.

2. Ezra Meeker to Harvey Scott, January 24, 1910, Meeker Papers, Box 7, Folder 10.

3. Ezra Meeker to Carrie Osborne, January 13, 1910, Meeker Papers, Box 7, Folder 10.

4. Ezra Meeker to Alden Blethen, March 5, 1910, Meeker Papers, Box 7, Folder 10.

5. Ezra Meeker to David Miller, February 14, 1910, Meeker Papers, Box 7, Folder 10.

6. Ezra Meeker to Edward Jay Allen, March 6, 1910, Meeker Papers, Box 7, Folder 10.

Notes for pages 60–64 231

7. R. H. Thompson to Ezra Meeker, March 9, 1910, Meeker Papers, Box 7, Folder 10.

8. J. M. Frink to Clarence Bagley, March 19, 1910, Meeker Papers, Box 7, Folder 10; H. W. Corbett, Jr., Willamette Iron Works to Ezra Meeker, March 9, 1910, Meeker Papers, Box 6, Vol. 8; and M. H. Broughton to Ezra Meeker, Meeker Papers, Box 5, Folder 22.

9. Ezra Meeker to Harvey Scott, January 24, 1910, Meeker Papers, Box 7, Folder 10.

10. Ezra Meeker to David Miller, February 14, 1910, Meeker Papers, Box 7, Folder 10.

11. Ezra Meeker to William Mardon, February 27, 1910, Meeker Papers, Box 6, Vol. 8.

12. *Journal of Ezra Meeker's 1910 Old Oregon Trail Monument Expedition*, March 21, 1910, Meeker Papers, Box 8, Folder 5E.

13. He described the location as four miles out from The Dalles on the public road, "and from which W. F. Cushing's house bears N 55 E about 200 rods. This house is down in the deep valley near 10 mile creek with a peach orchard in the foreground; took a photograph of this view four years ago. From the boulder another house near the road bears N 50 E about 100 rods distant." *Journal of Ezra Meeker's 1910 Old Oregon Trail Monument Expedition*, March 21, 1910, Meeker Papers, Box 8, Folder 5E. He also included a photograph of the boulder in his books *Ventures and Adventures of Ezra Meeker* (Seattle: Rainier Printing Co., 1909) and *The Busy Life of Eighty-Five Years of Ezra Meeker* (Indianapolis: Wm. B. Burford Press, 1916).

14. *Journal of Ezra Meeker's 1910 Old Oregon Trail Monument Expedition*, March 22, 1910, Meeker Papers, Box 8, Folder 5E.

15. *Journal of Ezra Meeker's 1910 Old Oregon Trail Monument Expedition*, March 21, 1910, Meeker Papers, Box 8, Folder 5E.

16. *Journal of Ezra Meeker's 1910 Old Oregon Trail Monument Expedition*, March 23–25, 1910, Meeker Papers, Box 8, Folder 5E. A picture of the Well Spring rock graced the cover of *Ventures and Adventures*.

17. A photograph of Cora Mardon leading the wagon being pulled by a team of oxen and mules was taken along this part of the drive. It was published in "Bridgeport Woman Tells of Ezra Meeker's Life," *Bridgeport (CT) Herald*, December 9, 1928, 3.

18. Roger Blair, "I Remember the Day: A Connection to Ezra Meeker and Oregon Trail History," *News from the Plains* (Independence, MO: Oregon-California Trails Association, January 1998), 1; "The Day Ezra Meeker Came to Our School," *Pioneer Trails*, Umatilla County Historical Society, Summer 1990, 14–16; "Ezra Meeker Visit Recalled," *The Times* (Waitsburg, WA), October 16, 1997, 3; "Sign to Mark Oregon Trail is in Works," *Milton-Freewater Valley Herald*, February 24, 1988, 3; and April 14, 1910 journal entry, Meeker Papers, Box 8, Folder 5E.

19. The Weston marker was placed in the road at O. M. Richards' place. The Athena marker was set at the crossing of Wild Horse Creek where the railroad crossed the trail and county roads.

20. Roger Blair and his wife Susan, OCTA members from Pendleton, Oregon, made a search for these markers using documentation supplied by the author and failed to find any.

21. Ezra Meeker to Carrie Osborne, May 6, 1910, Meeker Papers, Box 7, Folder 10.

22. Ezra Meeker to Senator Piles, April 23, 1910, Meeker Papers, Box 7, Folder 10.

23. *Journal of Ezra Meeker's 1910 Old Oregon Trail Monument Expedition*, April 23, 1910, Meeker Papers, Box 8, Folder 5E.

24. Ezra Meeker to Dr. William T. Phy, May 1, 1910, Meeker Papers, Box 6, Vol. 9.

25. William Mardon to Ezra Meeker, May 6, May 9, and May 13, 1910, Meeker Papers, Box 7, Folder 10.

26. William Mardon to Ezra Meeker, May 6, 1910, Meeker Papers, Box 7, Folder 10.

27. Ezra Meeker to William Mardon May 7, 1910, Meeker Papers, Box 6, Vol. 9.

28. www.theodore-roosevelt.com/trspeechescomplete.html.

29. *Journal of Ezra Meeker's 1910 Old Oregon Trail Monument Expedition*, May 15, 1910, Meeker Papers, Box 8, Folder 5E.

30. Ezra Meeker to Carrie Osborne, May 9, 1910, and May 14, 1910, Meeker Papers, Box 6, Vol. 9.

31. Ibid.

32. Ezra Meeker to William Mardon, May 15, 1910, Meeker Papers, Box 6, Vol. 9.

33. *Journal of Ezra Meeker's 1910 Old Oregon Trail Monument Expedition*, May 15, 1910, Meeker Papers, Box 8, Folder 5E.

34. Ibid.

35. Ezra Meeker to Joe Templeton, May 21 and May 24, 1910, Meeker Papers, Box 6, Vol. 9.

36. Moody came west over the Oregon Trail in 1855 as a child, ran a mercantile business and served as a U.S. Representative from 1899 to 1903.

37. Also spelled Utter, the Otter Massacre of September 1860 was a running battle between a party of forty-four immigrants headed by Elijah P. Utter and the Shoshone/Bannock Indians. Thirty-two immigrants died in the battle, which is marked at a number of locations.

38. Ezra Meeker to Mrs. D. T. Standrod, July 11, 1910, Meeker Papers, Box 6, Vol. 10.

39. Ezra Meeker to William Mardon, July 9, 1910, Meeker Papers, Box 6, Vol. 10.

40. William Mardon to Ezra Meeker, July 18, 1910, Meeker Papers, Box 7, Folder 10.

41. The details of Frank Meeker and his misdeeds may be found in Larsen, *Hop King*, 153–58.

42. Ezra Meeker to Carrie Osborne, July 21, 1910, Meeker Papers, Box 6, Vol. 10.

43. Joe Templeton to Ezra Meeker, August 15, 1910, Meeker Papers, Box 7, Folder 10.

44. Ezra Meeker to Carrie Osborne, August 22, 1910, Meeker Papers, Box 6, Vol. 10.

45. William Mardon to Ezra Meeker, September 27, 1910, Meeker Papers, Box 7, Folder 10. The maps never made it to Congress. Today they are stored in Box 16 of

the Meeker Papers at the Washington State Historical Society Research Center in Tacoma, Washington.

46. George Martin to Jesse Palmer, September 12, 1910, Meeker Papers, Box 7, Folder 10. Despite this glowing recommendation, Meeker failed to get into the fair.

47. William Mardon to Ezra Meeker, September 27, 1910, Meeker Papers, Box 7, Folder 10.

48. Ezra Meeker to William Mardon, October 1, 1910, Meeker Papers, Box 7, Folder 10.

Chapter 7: The Second Expedition, 1911

1. Ezra Meeker to W. E. Humphrey, January 3, 1911, Meeker Papers, Box 6, Vol. 12.

2. Ezra Meeker to W. E. Humphrey, January 13, 1911, Meeker Papers, Box 6, Vol. 12.

3. W. E. Humphrey to Ezra Meeker, January 14, 1911, Meeker Papers, Box 7, Folder 11.

4. Ezra Meeker to W. E. Humphrey, January 21, 1911, Meeker Papers, Box 6, Vol. 12.

5. Ezra Meeker to Columbus Lecture Bureau, January 5, 1911, Meeker Papers, Box 6, Vol. 12.

6. Ezra Meeker to Chief of Police, Dayton, Ohio, January 26, 1911, Meeker Papers, Box 6, Vol. 14.

7. Ezra Meeker to Carrie Osborne, February 9, and February 18, 1911, Meeker Papers, Box 6, Vol. 13.

8. Ezra Meeker to M. B. Golden, February 27, 1911, Meeker Papers, Box 6, Vol. 13.

9. Ezra Meeker to Carrie Osborne, February 28, 1911, Meeker Papers, Box 6, Vol. 13.

10. Ezra Meeker to Carrie Osborne, March 13, 1911, Meeker Papers, Box 6, Vol.13.

11. "Open Letter to the Mayor," Meeker Papers, Box 1, Folder 4.

12. Ezra Meeker to Carrie Osborne, May 31, 1911, Meeker Papers, Box 6, Folder 14.

13. Ezra Meeker to Mayor, Rochester, NY, June 20, 1911, Meeker Papers, Box 6, Vol. 14.

14. The document was dated June 1, but it came with a June 12 letter from his daughter, Carrie Osborne. Carrie and Eben Osborne to Ezra Meeker, June 1, 1911, Meeker Papers, Box 4, Folder 3A.

15. Ezra Meeker to Eben Osborne, June 20, 1911, Meeker Papers, Box 6, Vol. 14.

16. While it seems that Meeker was writing sarcastically, he did, in the end, outlive Eben. Ezra Meeker to Carrie and Eben Osborne June 19, 1911, Meeker Papers, Box 6, Vol. 14.

17. Ezra Meeker to John Hartman Jr., September 6, 1911, Meeker Papers, Box 6, Vol. 14.

18. Ezra Meeker to Carrie Osborne, August 31, 1911, Meeker Papers, Box 6, Vol. 14.

19. Ezra Meeker to Carrie Osborne, September 18, 1911, Meeker Papers, Box 6, Vol. 15.

20. Ezra Meeker to Carrie Osborne, October 9, 1911, Meeker Papers, Box 6, Vol. 15.

21. Ezra Meeker to Joe Templeton, November 17, 1911, Meeker Papers, Box 6, Vol. 15.

22. Ezra Meeker to Joe Templeton, February 6, 1912, Meeker Papers, Box 6, Vol. 16.

23. Joe Templeton to Ezra Meeker, February 12, 1912, Meeker Papers, Box 7, Folder 12.

24. A. R. Corey to Ezra Meeker, July 28, 1911, Meeker Papers, Box 7, Folder 11.

25. Ezra Meeker to Secretary Cotton Palace Fair, October 10, 1911, Meeker Papers, Box 6, Vol. 15.

26. Ezra Meeker to Nichols, Dean & Gregg, October 4, 1911, Meeker Papers, Box 6, Vol. 15. Nichols was a former trustee of Meeker's Puyallup Hop Company.

27. Ezra Meeker to Joe Grimes, October 19, 1911, Meeker Papers, Box 6, Vol. 15.

28. Ezra Meeker to Curt Teich & Co., November 30, 1911, Meeker Papers, Box 6, Vol. 15.

29. Ezra Meeker to William Mardon, October 22, 1911, Meeker Papers, Box 6, Vol. 15.

30. Ezra Meeker to Professor Cantwell, November 5, 1911, Meeker Papers, Box 6, Vol. 15.

31. Ezra Meeker to Joe Grimes, November 12, 1911, Meeker Papers, Box 6, Vol. 15.

32. Ezra Meeker to Billy Boy, December 25, 1911, Meeker Papers, Box 6, Vol. 16.

Chapter 8: The Second Expedition, 1912

1. W. E. Humphrey to Ezra Meeker, January 18, 1912, Meeker Papers, Box 7, Folder 12.

2. Ezra Meeker to W. S. McAdoo & Co., March 1, 1912, Meeker Papers, Box 6, Vol. 16.

3. Ezra Meeker to Carrie Osborne, March 4, 1912, Meeker Papers, Box 7, Folder 22.

4. Ezra Meeker to Curt Teich & Co., March 8, 1912, Meeker Papers, Box 6, Vol. 16.

5. Curt Teich & Co. to Ezra Meeker, March 12, 1912 Meeker Papers, Box 6, Vol. 16.

6. Ezra Meeker to W. S. McAdoo & Co., March 20, 1912, Meeker Papers, Box 6, Vol. 16.

7. Ezra Meeker to Carrie Osborne, April 12, 1912, Meeker Papers, Box 6, Vol. 17.

8. Ezra Meeker to William Mardon, April 10, 1912, Meeker Papers, Box 6, Vol. 14.

9. Mrs. Ethel Miner to Ezra Meeker, June 10, 1912, Meeker Papers, Box 7, Folder 12.

10. Ezra Meeker to Mrs. Ethel Miner, June 19, 1912, Meeker Papers, Box 6, Vol. 17.

11. Ezra Meeker to William Mardon, April 20 and May 1, 1912, Meeker Papers, Box 6, Vol. 17.

12. Mrs. Ethel Miner to Ezra Meeker, February 5, 1914, Meeker Papers, Box 7, Folder 14–15.

13. "Bridgeport Woman Tells of Ezra Meeker's Life," *The Bridgeport Herald*, December 9, 1928, 3; "Covered Wagon Traveler Dies," *Bridgeport Times-Star*, December 9, 1928, 3. See www.findagrave.com for more information on the Mardons.

14. Ezra Meeker to Carrie Osborne, April 29, 1912, Meeker Papers, Box 6, Vol. 17.

15. Meeker never mentioned Mrs. Barrett's first name. Ezra Meeker to Carrie Osborne, May 2, 1912, Meeker Papers, Box 6, Vol. 17.

16. Ezra Meeker to Carrie Osborne, May 2, 1912, Meeker Papers, Box 6, Vol. 17.

17. Ezra Meeker to Carrie Osborne, April 27, 1912, Meeker Papers, Box 6, Vol. 17.

18. Ezra Meeker to the Wright brothers, February 1, 1911, Meeker Papers, Box 6, Vol. 13.

19. Ezra Meeker to Cash Register Co., April 12, 1911, Meeker Papers, Box 6., Vol. 13.

20. Ezra Meeker to Rock Island R.R. Lost Property Manager, March 23, 1912, Meeker Papers, Box 6, Vol. 16.

21. Meeker does not explain how or where the film canister was found. However, in an April 12, 1912, letter written in Omaha to his daughter Meeker wrote, "My moving pictures with the ox team in Dayton and Toledo has been a drawing card." Ezra Meeker to Carrie Osborne, April 12, 1912, Meeker Papers, Box 6, Vol. 17.

22. Ezra Meeker to Joe Grimes, April 3, 1912, Meeker Papers, Box 6, Vol. 17.

23. Ibid.

24. In addition to promoting Wild West shows, Irwin was a lobbyist for the Union Pacific and president of the Cheyenne Feature Film Co.

25. Ezra Meeker to C. B. Irwin, April 21, 1912, Meeker Papers, Box 6, Vol. 17.

26. Ezra Meeker to Carrie Osborne, April 4, 1912, Meeker Papers, Box 6, Vol. 17.

27. Ezra Meeker to William Mardon, April 20, 1912, Meeker Papers, Box 6, Vol. 17.

28. Ezra Meeker to Carrie Osborne, April 27, 1912, Meeker Papers, Box 6, Vol. 17.

29. Ezra Meeker to Carrie Osborne, May 2, 1912, Meeker Papers, Box 6, Vol. 17.

30. Ezra Meeker to Oscar A. Albrecht, May 19, 1912, Meeker Papers, Box 6, Vol. 17.

31. Ezra Meeker to Carrie Osborne, May 24, 1912, Meeker Papers, Box 6, Vol. 17.

32. Meeker's estranged nephew Frank Meeker lived in Fort Morgan for a time and Frank's wife was buried there. Frank was remarried and living in Oregon in 1912. For the story of the estrangement see Larsen, *Hop King*, 141–58. Ezra Meeker to Oscar A. Albrecht, July 2, 1912, Meeker Papers, Box 6, Vol. 17.

33. Oscar A. Albrecht to Ezra Meeker, July 16, 1912, Meeker Papers, Box 5, Folder 22.

34. Ezra Meeker to O. A. Albrecht, July 20, 1912, Meeker Papers, Box 6, Vol. 17.

35. Ezra Meeker to Carrie Osborne, August 9, 1912, Meeker Papers, Box 6, Vol. 18.

36. Ezra Meeker to P. W. Buckwalter, July 24, 1912, Meeker Papers, Box 6, Vol. 18.

37. "Ezra Meeker Lashes Busybody," *Cowlitz County Advocate* (Castle Rock, WA), August 8, 1912, 4; "The Round-Up," *Daily Capital Journal* (Salem, OR), August 5, 1912, 2; "Meeker Plies Whip," *Ashland Tidings*, August 8, 1912, 1.

38. Ezra Meeker to Carrie Osborne, August 9, 1912, Meeker Papers, Box 6, Vol. 18.

236 Notes for pages 103–110

39. Ezra Meeker to Charles Hood, August 14, 1912, Meeker Papers, Box 6, Vol. 18.

40. W. S. McAdoo & Co. to Ezra Meeker, August 21, 1912, Box 5, Folder 22.

41. Ezra Meeker to President of Montpelier, Idaho, Women's Club, December 4, 1912, Meeker Papers, Box 6, Vol. 18.

42. Senator Samuel H. Piles retired from the Senate in 1911 and was replaced by Miles Poindexter. Senator Wesley L. Jones remained in office from 1905 to 1932.

43. Ezra Meeker to W. E. Humphrey, undated, Meeker Papers, Box 6, Vol. 18.

44. Ezra Meeker to John P. Hartman Jr., December 28, 1912, Meeker Papers, Box 6, Vol. 18.

CHAPTER 9: INTERLUDE

1. "Meeker Pleads for Pioneers," *Puyallup Tribune*, September 28, 1912, 18.

2. Ezra Meeker to President, Puyallup Ladies Club, October 25, 1912, Meeker Papers, Box 6, Vol. 18.

3. "Favor Change Seven to One," *Puyallup Tribune*, November 16, 1912, 18.

4. "Ezra Meeker's Book," *Seattle Post-Intelligencer*, December 15, 1912, 18.

5. Ezra Meeker to Senator Wesley L. Jones, December 15, 1912, Meeker Papers, Box 6, Vol. 18.

6. Ezra Meeker to H. W. Meyers, October 27, 1912, Meeker Papers, Box 4, Folder 8.

7. Meeker Papers, Box 4, Folder 8.

8. "The Last Drive," *Tacoma Daily Ledger*, February 23, 1913, 9.

9. M. Mayer, Secretary Metropolitan Park Board to Ezra Meeker, June 6, 1913, Meeker Papers, Box 4, Folder 8.

10. Ezra Meeker to Mrs. H. B. Patton, October 12, 1913, Meeker Papers, Box 6, Vol. 21.

11. Ezra Meeker to Richard Bath, November 16, 1913, Meeker Papers, Box 6, Vol. 21.

12. George Lewis Gower to Ezra Meeker, June 12, 1914, Meeker Papers, Box 4, Folder 8.

13. Ezra Meeker to the Park Board, June 16, 1914, Meeker Papers, Box 6, Vol. 22.

14. Ezra Meeker to the Park Board, July 2, 1914, Meeker Papers, Box 6, Vol. 22.

15. Ezra Meeker to Frederick Heath, President of the Park Board, November 5, 1914, and November 13, 1914, Meeker Papers, Box 6, Vol. 23.

16. Ezra Meeker to Ethel Miner, January 25, 1914, Meeker Papers, Box 6, Vol. 21.

17. For the full story of this train trip see Larsen, *Hop King*, 51.

18. Ezra Meeker to Olive Osborne, December 5, 1913, Meeker Papers, Box 4, Folder 3A.

19. Mrs. Ethel Miner to Ezra Meeker, February 5, 1914, Meeker Papers, Box 7, Folder 14–15.

20. Ezra Meeker to Jo [Pense Meeker], March 29, 1914, Meeker Papers, Box 6, Vol. 22.

Notes for pages 111–120 237

21. Ezra Meeker, *Seventy Years of Progress in Washington* (Tacoma, WA: Allstrum Printing Company, 1921), 52. See also Larsen, *Hop King*, 37–40.

22. "Fifty Girls Will Go to Rainier," *Seattle Post-Intelligencer*, March 15, 1914, 8; Camp Fire Girls was created as a nonsectarian organization for girls in 1912. It was intended to be the sister to the Boy Scouts of America with programs emphasizing camping and other outdoor activities. By 1913 its membership was 60,000.

23. Ezra Meeker to Charles Moore, December 2, 1914, Meeker Papers, Box 6, Vol. 23.

24. Ezra Meeker to Richard Seeley Jones, March 25 and April 1, 1915, Meeker Papers, Box 6, Vol. 24.

25. "Meeker to Start New Ox Team Trip," *Seattle Post-Intelligencer*, May 4, 1914, 2.

26. Ezra Meeker to H. A. Kennedy, May 22, 1914, Meeker Papers, Box 6, Vol. 22.

27. Ezra Meeker to Carrie Osborne, August 18, 1914, Meeker Papers, Box 6, Vol. 23.

28. Ezra Meeker to Richard Seeley Jones, August 15, 1914, Meeker Papers, Box 6, Vol. 23.

29. Ezra Meeker to Carrie Osborne, August 17, 1914, Meeker Papers, Box 6, Vol. 23.

30. Ezra Meeker to Governor Lister, September 30, 1914, Meeker Papers, Box 6, Vol. 23.

31. Ezra Meeker to Governor Lister, October 14, 1914, Meeker Papers, Box 6, Vol. 23.

32. Ezra Meeker to Curt Teich & Co., November 2, 1914, Meeker Papers, Box 6, Vol. 23.

33. Ezra Meeker to Richard Seeley Jones, September 10, 1914, Meeker Papers, Box 6, Vol. 23.

34. Ezra Meeker to Frederick Heath, November 13, 1914, Meeker Papers, Box 6, Vol. 23.

35. Ezra Meeker to Richard Seeley Jones, October 7, 1914, Meeker Papers, Box 6, Vol. 23.

36. Ezra Meeker to Richard Seeley Jones, October 24, 1914, Meeker Papers, Box 6, Vol. 23.

37. Ezra Meeker to Governor Lister, November 24, 1914, Meeker Papers, Box 6, Vol. 23.

38. Ezra Meeker to Governor Lister, December 18, 1914, Meeker Papers, Box 6, Vol. 23.

39. Ezra Meeker to Grace Raymond Hebard, December 19, 1914, Meeker Papers, Box 6, Vol. 23; Ezra Meeker to Richard Seeley Jones, December 24, 1914, Meeker Papers, Box 6, Vol. 23.

40. Meeker's Address, January 21, 1919, Meeker Papers, Box 4, Folder 8.

Chapter 10: The Pathfinder Expedition, Part 1: 1915–1916

1. John M. Studebaker to Ezra Meeker, October 20, 1915, Meeker Papers, Box 7, Folder 16–17B.

2. Ezra Meeker to John M. Studebaker, November 4, 1915, Meeker, Papers, Box 6, Vol. 24.

3. John M. Studebaker to Ezra Meeker, November 29, 1915, Meeker Papers, Box 7, Folder 16–17B.

4. "Ezra Meeker, 85, Celebrates at National Capital," *Tacoma, News*, December 28, 1915, 1.

5. An advertisement for the Pathfinder automobile was inserted into *The Busy Life of Eighty-Five Years of Ezra Meeker*.

6. "Story of a Trip from Washington to Washington," Meeker Papers, Box 8, Folder 2–3. January 9, 1917 entry. Meeker's confession came in this unpublished story.

7. Woodrow Wilson was the third of four presidents that Meeker would personally meet: Benjamin Harrison, Theodore Roosevelt (at least twice), Woodrow Wilson, and Calvin Coolidge. He corresponded with future President Herbert Hoover during World War I about food relief. A summary of Meeker's meeting with President Wilson is contained in Ezra Meeker to F. G. Buskirk, February 14, 1916, Meeker Papers, Box 7, Folder 18–26A; Ezra Meeker to Carrie Osborne, February 14, 1916, Meeker Papers, Box 7, Folder 18–26A; and Ezra Meeker to William Bonney, February 15, 1916, Meeker Papers, Box 7, Folder 18–26A. "Meeker Visits President for Oregon Trail Bill," *Seattle Post-Intelligencer*, February 15, 1916, 1, and "Ezra Meeker Discusses Trails with President," *Seattle Times*, February 14, 1916, 4, give a brief description of Meeker's visit with President Wilson.

8. The entire statement that Meeker gave the president is contained in a letter from Ezra Meeker to F. G. Buskirk, February 14, 1916, Meeker Papers, Box 7, Folder 18–26A.

9. Ezra Meeker to Carrie Osborne, February 14, 1916, Meeker Papers, Box 7, Folder 18–26A.

10. Ezra Meeker to William Bonney, February 19, 1916, Meeker Papers, Box 7, Folder 18–26A.

11. Carrie Osborne to Ezra Meeker, April 3, 1916, Meeker Papers, Box 4, Folder 4B.

12. F. G. Buskirk to Ezra Meeker, April 29, 1916, Box 7, Folder 18–26C and F. G. Buskirk to Ezra Meeker, May 15, 1916, Meeker Papers, Box 7, Folder 18–26D.

13. Carrie Osborne to Ezra Meeker, May 18, 1916, Meeker Papers, Box 4, Folder 4B.

14. F. G. Buskirk to Ezra Meeker, May 16, 1916, Meeker Papers, Box 7, Folder 18–26D.

15. Ezra Meeker to Earl Schaffer, May 24, 1916, Meeker Papers, Box 7, Folder 18–26D.

16. William E. Humphrey to Ezra Meeker, June 3, 1916, Meeker Papers, Box 7, Folder 18–26D.

17. William Mardon to Ezra Meeker, June 6, 1916, Meeker Papers, Box 7, Folder 18–26D.

18. F. G. Buskirk to Ezra Meeker, June 21, 1916, Meeker Papers, Box 7, Folder 18–26D.

19. F. G. Buskirk to Ezra Meeker, June 29, 1916, Meeker Papers, Box 7, Folder 18-26D.

20. F. G. Buskirk to Ezra Meeker, July 1, 1916, Meeker Papers, Box 7, Folder 18-26E.

21. Miss Pendergast to Ezra Meeker, July 5, 1916, Meeker Papers, Box 7, Folder 18-26E.

22. F. G. Buskirk to Ezra Meeker, July 15, 1916, Meeker Papers, Box 7, Folder 18-26E.

23. F. G. Buskirk to Ezra Meeker, July 22, 1916, Meeker Papers, Box 7, Folder 18-26E.

24. Ezra Meeker's Day Calendar for 1916 supplied a day-to-day list of locations and gives some detail of his activities. Meeker Papers, Box 8, Folder 5G.

25. Day Calendar, Meeker Papers, Box 8, Folder 5G.

26. *Topeka Capital Journal*, July 29, 1916, Meeker Papers, Box 14, Folder 6.

27. F. G. Buskirk to Ezra Meeker August 15, 1916, Meeker Papers, Box 7, Folder 18-26E.

28. F. G. Buskirk to Ezra Meeker August 10, 1916, Meeker Papers, Box 7, Folder 18-26E.

29. "Ezra Meeker in Lander," *Wind River Mountaineer* (Lander, WY), August 18, 1916, 1.

30. Tacoma Speech by Ezra Meeker, October 12, 1916, Meeker Papers, Box 8, Envelope 3. The stagecoach today resides in the basement of the Washington State Historical Society Research Center in Tacoma, Washington. The current curators do not seem to prize it as highly as did Meeker and Bonney. When viewing it, the curator asked the author if he knew anyone who was in need of a stagecoach.

31. There is a photograph of Meeker, the Pathfinder, and Hermon Nickerson taken in Lander in the author's collection and in the Lander Museum. Working with a $4,000 state appropriation, Nickerson succeeded in erecting nineteen trail markers in Wyoming. The two men were definitely kindred spirits. Six years later Walter Meacham of Baker City, Oregon, joined the cast of monument builders with the creation of the Old Oregon Trail Association. Apparently Meeker's passion for monuments was contagious.

32. "Ezra Meeker Again Visits Lander, On His Trip West," *Wyoming State Journal*, August 18, 1916, 1. "Ezra Meeker in Lander," *Wind River Mountaineer*, August 18, 1916, 1.

33. A copy of the pamphlet may be found in the Meeker Papers, Box 8, Folder 4.

34. Standrod also organized the Red Cross for Bannock County, Idaho.

35. Minnie Francis Howard and her husband were instrumental in founding the Pocatello General Hospital in 1907. She became interested in the history of Fort Hall after Meeker's 1906 visit to Pocatello. "Dr. Minnie Howard," *Overland Journal*, Summer 2017, 80.

36. D. K. Hall, Director of Services, the Pathfinder Company to Ezra Meeker, September 9, 1916, Meeker Papers, Box 7, Folder 18-26E.

37. "Oregon Trail Pioneer Here," *Idaho Statesman*, September 4, 1916, 5.

38. "Ezra Meeker Arrives on Continental Trip," Meany Pioneer Files, Edmund S. Meany Papers, Special Collections, University of Washington Library.

39. "Ezra Meeker Returns in Schooner Run by Gasoline," *Puyallup Valley Tribune*, September 16, 1916, 1.

40. "Valley Pioneer Meets His Friends at the Fair, Ezra Meeker and His Schoonermobile Attract Much Attention After Long Transcontinental Tour," *Puyallup Valley Tribune*, September 23, 1916.

41. "Ezra Meeker," *Ellensburg Capital*, September 28, 1916, 3.

42. "Gathers Historical Data in Yakima County," *Yakima Daily Republic*, September 25, 1916, 2.

43. "Ezra Meeker Pays Visit to City," *Bremerton Searchlight*, September 29, 1916, 3.

44. Ezra Meeker to the Pathfinder Company, October 2, 1916, Meeker Papers, Box 7, Folder 18–26E.

45. In 1906 Meeker attempted to place markers at a number of sites along the Cowlitz Trail (the Washington segment of the Oregon Trail). He succeeded only in Tenino. In 1916 the Daughters and Sons of the American Revolution picked up the baton and put up fourteen Oregon Trail markers along the Cowlitz Trail from Tumwater to Vancouver.

46. Willie D. Pendergast to Ezra Meeker, October 4, 1916, Meeker Papers, Box 7, Folder 18–26E.

47. "Ezra Meeker Will Tell of His Trip Across Continent on 'Pioneer Way' at Meeting This Evening," *Everett Herald*, October 6, 1916, 7.

48. Ezra Meeker to Pathfinder Company, October 6, 1916, Meeker Papers, Box 7, Folder 18–26E.

49. F. C. Buskirk to Ezra Meeker, October 6, 1916, Meeker Papers, Box 7, Folder 18–26E.

50. "Meeker Pleads for Highway Aid," *Tacoma Ledger*, October 15, 1916, 18.

Chapter 11: The Pathfinder Expedition, Part 2: 1916

1. This bill would merge the old National Road with the Oregon Trail, thus creating a true transcontinental highway.

2. Ezra Meeker to Charles Davis, October 9, 1916, Meeker Papers, Box 7, Folder 18–26E.

3. "Trail Blazer in Centralia," *Centralia Chronicle*, October 14, 1916, 1.

4. "Ezra Meeker Visits," *Chehalis Bee Nugget*, October 20, 1916, 2.

5. "E. Meeker on Trail," *Oregonian*, Monday, October 16, 1916, 5.

6. Ezra Meeker to Carrie Osborne, October 18, 1916, Meeker Papers, Box 7, Folder 18–26E.

7. Ezra Meeker to Pathfinder Company, October 18, 1916, Meeker Papers, Box 7, Folder 18–26E.

8. Ezra Meeker to J. G. Buskirk, October 25, 1916, Meeker Papers, Box 7, Folder 18–26E.

9. Ezra Meeker to Charles Davis, October 27, 1916, Meeker Papers, Box 7, Folder 18–26E.

10. Ezra Meeker to William Bonney, November 6, 1914, Meeker Papers, Box 7, Folder 18–26C.

11. Ezra Meeker to George Lufkin Co., agent and distributor of the Pathfinder car at Los Angeles for Southern California and Arizona, October 31, 1916, Meeker Papers, Box 7, Folder 18–26E.

12. Ezra Meeker to Charles Davis, November 6, 1916, Meeker Papers, Box 7, Folder 18–26C.

13. Carrie Osborne to Ezra Meeker, November 10, 1916, Meeker Papers, Box 4, Folder 4B.

14. Ezra Meeker to Pathfinder Company, November 12, 1916, Meeker Papers, Box 7, Folder 18–26C.

15. Marion Meeker to Ezra Meeker, November 11, 1916, Meeker Papers, Box 4, Folder 4B.

16. Meeker's correspondence from this time in southern California is almost nonexistent. There is only one letter in December 1916 from his daughter and no business correspondence. With the assistance of Ray Kanter, Marion Meeker's great-grandson, the author made a canvass of southern California libraries and possible repositories of Meeker materials, locating a series of newspaper articles and photographs that shed some light on Meeker's activities.

17. "Trail Blazer Will Meet Pioneers," *San Bernardino Sun*, December 2, 1916.

18. "Ezra Meeker Addressed the Old Pioneers in Log Cabin," *San Bernardino Sun*, December 3, 1916.

19. "Pioneer Trail Blazer Urges Southern National Highway," *San Diego Union*, December 7, 1916.

20. "Young-old Pioneer Retracing Steps Over Old Trail," *Long Beach Daily Telegram*, December 12, 1916.

21. "Meeker Urges Approval of Chamber," *San Bernardino Sun*, December 15, 1916.

22. "Ezra Meeker Is Here Today," *Redlands Daily Facts*, December 15, 1916, 1.

23. Carrie Osborne to Ezra Meeker, December 20, 1916, Meeker Papers, Box 4, Folder 4C.

Chapter 12: The Pathfinder Expedition, Part 3: 1917

1. The brothers signed their name Johnstone on the contract. Meeker spelled it Johnson in his manuscript. Meeker Papers, Box 7, Folder 30.

2. "Story of a Trip from Washington to Washington," Meeker Papers, Box 8, Folder 2–3.

3. Meeker's 1917 Day Calendar, Meeker Papers, Box 11.

4. Ezra Meeker to Newton Baker, April 6, 1917, Meeker Papers, Box 7, Folder 27A.

5. Ezra Meeker to Miles Poindexter, April 11, 1917, Meeker Papers, Box 7, Folder 27A.

6. 1917 Day Calendar, Meeker Papers, Box 11, Envelope 3, Folder 3. Ezra Meeker to William Bonney, April 14, 1917, Meeker Papers, Box 4, Folder 7B.

7. William Bonney to Senator J. Hamilton Lewis, April 5, 1917, Meeker Papers, Box 4, Folder 7B.

8. Eben Osborne to Ezra Meeker, April 18, 1917, Meeker Papers, Box 4, Folder 4C.

Chapter 13: World War I

1. Ezra Meeker to William Bonney, April 22, 1917, Meeker Papers, Box 4, Folder 7B.

2. Ezra Meeker to the Luther Manufacturing Company, June 1, 1917, Meeker Papers, Box 6, Vol. 24 and Ezra Meeker to Professor J. S. Caldwell, June 1, 1917, Meeker Papers, Box 6, Vol. 24.

3. Ezra Meeker to the Wisconsin Fiber Container Corporation, June 6, 1917, Meeker Papers, Box 6, Vol. 24.

4. "Big Hop Crop Expected," *Puyallup Valley Tribune* August 18, 1916, 1; Larsen, *Hop King*, 199–202.

5. Herbert Hoover, U.S. president 1929–33, became internationally known for his efforts in European food relief from 1914 to 1916. President Wilson appointed him to lead the U.S. Food Administration during World War I.

6. Ezra Meeker to the Wisconsin Fiber Container Corporation, June 6, 1917, Meeker Papers, Box 6, Vol. 24.

7. Ezra Meeker to the Wittenberg-King Co., June 14, 1917, Meeker Papers, Box 6, Vol. 24.

8. 1917 Day Calendar, June 20 and 22, 1917, Meeker Papers, Box 11, Envelope 3, Folder 3.

9. Ezra Meeker to Charles Lilly, August 14, 1917, Meeker Papers, Box 6, Vol. 24.

10. Ezra Meeker to Charles Hood, July 25, 1917, Meeker Papers, Box 6, Vol. 24.

11. Ezra Meeker to Herbert Hoover, August 12, 1917, Meeker Papers, Box 6, Vol. 24.

12. 1917 Day Calendar, August 9, 1917, Meeker Papers, Box 11, Envelope 3, Folder 3.

13. "Ezra Meeker on Pioneer Life," *Arlington Times*, August 9, 1917, 1.

14. Ezra Meeker to Bertha Templeton, December 13, 1917, Meeker Papers, Box 4, Folder 4C.

15. William Bonney to Ezra Meeker, April 27, 1918, Meeker Papers, Box 4, Folder 7B. 1919 Day Calendar. Meeker Papers, Box 11, Envelope 3, Folder 4.

16. Olive Osborne to Ezra Meeker, February 16, 1918, Meeker Papers, Box 4, Folder 4C.

17. "Ezra Meeker's Garden Yields $257.80 to Aid in War Work of Red Cross, Pioneer Who Followed Oregon Trail Shows He's Not Too Old for Labor If He Is Far Over Age for Service," *Seattle Post-Intelligencer*, November 24, 1918, 27.

18. Ezra Meeker to Charles Hebbard, March 25, 1918, Meeker Papers, Box 6, Vol. 25.

19. Ezra Meeker to Senator Wesley Jones, March 25, 1918, Meeker Papers, Box 6, Vol. 25.

20. Ezra Meeker to Olive Osborne, December 13, 1918, and January 3, 1919, Meeker Papers, Box 6, Vol. 26.

21. "League of Nations," Meeker Papers, Box 10, Envelope 2.

Chapter 14: The Fight for Naches Pass

1. The story of the building of the Naches Pass wagon road is told in Karen L. Johnson and Dennis M. Larsen, *A Yankee on Puget Sound: Pioneer Dispatches of Edward Jay Allen, 1852–1855* (Pullman: Washington State University Press, 2013).

2. Ray Egan, "On the Trail of a Legend: The Ox Rope Story," *Overland Journal* 34, no. 2 (Summer 2016), 58–69; and Meeker and Driggs, *Ox-Team Days on the Oregon Trail*, 118.

3. Elva Cooper Magnusson, "The Naches Pass," *Washington Historical Quarterly* 25 (July 1934), 163–70.

4. George Himes to Ezra Meeker, April 24, 1911, Meeker Papers, Box 4, Folder 12B.

5. George Himes to Ezra Meeker, July 7, 1919, Meeker Papers, Box 4, Folder 12B.

6. George Himes to Ezra Meeker, July 9, 1919, Meeker Papers, Box 4, Folder 12B.

7. "Party of Pioneers Stops in Olympia," *Washington Standard*, July 18, 1919. 1. "Ezra Meeker Issues Call to All Early-Day Trail Blazers," *Seattle Post-Intelligencer*, June 8, 1919, 26.

8. "Ninety Years No Bar to Activity," *Yakima Herald*, undated clipping, Meeker Papers, Box 14, Folder 6.

9. "Ezra Meeker Retraces His Steps Over Naches Trail," *Yakima Herald*, July 19, 1920, Meeker Papers, Box 14, Folder 6.

10. According to the July 29, 1920, entry in Meeker's day calendar Ranger Albro found the watch and returned it to a grateful Ezra. The watch was likely purchased by Meeker's wife from Tiffany's of New York as a gift to her husband on his 59th birthday. His intent was to pass it on to his eldest male grandchild. Meeker Papers, Box 11, Envelope 3, Folder 4.

11. "Pioneer, Ninety, Repeats Naches Pass Trip of Youth," *Seattle Post-Intelligencer*, July 19, 1920, 2.

12. This pass was at times spelled Carlton in Meeker's writings.

13. Ezra Meeker, "Voice of the People," *Centralia Chronicle*, July 28, 1920, 2.

14. "Ezra Meeker In Resentful Mood," *Aberdeen World*, July 27, 1920, 2.

15. Ezra Meeker to an unnamed legislator, January 3, 1921, Meeker Papers, Box 5, 31A.

16. Ezra Meeker, "The Trans-Cascade Highway," *Seattle Post-Intelligencer*, January 28, 1921, 6.

17. Ezra Meeker to Howard Driggs, February 1, 1921, Driggs Collection, B39F14.

18. Ezra Meeker, "The Naches Pathfinders," *Seattle Post-Intelligencer*, February 6, 1921, 5.

19. Ezra Meeker to David Longmire, March 14, 1921, Meeker Papers, Box 5, Folder 31A.

20. "Wants Road Over Old Naches Trail, Ezra Meeker Lays His Case Before Enumclaw Commercial Club Monday Evening," *Enumclaw Herald*, August 12, 1921, Meeker Papers, Box 14, Folder 6.

21. "Famous Oregon Trail Is Favored By Ezra Meeker. Pioneer Trail Blazer Says There is No Other Route West from Missouri River to Compare With It in Directness, Grade and Practical Use," November 2, 1921, Meeker Papers, Box 14, Folder 6.

22. "Meeker Boosts for Naches Pass Road," *Yakima Daily Republic*, March 8, 1922, 3.

23. Walter F. Tuesley and W. D. Robinson to Ezra Meeker, May 29, 1922 and G. C. Finley to Ezra Meeker, June 3, 1922, Meeker Papers, Box 7, Folder 27B.

24. Ezra Meeker to William Bonney, November 19, 1922, Meeker Papers, Box 4 Folder 7C.

25. Ezra Meeker to William Bonney, December 21, 1922, Meeker Papers, Box 4, Folder 7C.

26. "Meeker Sees Deception in Pass Report," *Seattle Post-Intelligencer*, June 19, 1924, 1.

27. "Meeker Road Claim Denied," *Seattle Post-Intelligencer*, June 20, 1924, 4.

28. "Meeker See Deception in Pass Proposal," *Seattle Post-Intelligencer*, June 19, 1924, 1 and "Highway on Wrong Route Says Meeker," *Seattle Post-Intelligencer*, June 21, 1924, 12.

29. "Ezra Meeker at 94 Will Run for Office," *New York Times*, July 11, 1924, 6.

30. "The State Needs the Naches Pass Highway," Meeker Papers, Box 8, Folder 4.

31. "Ezra Meeker, 'Trail Blazer' to Make Airplane Hop for Political Speech," *Seattle Post-Intelligencer*, August 11, 1924, Part 2, 1.

32. The author has documented five airplane trips by Meeker. But Ezra claimed in October 1924, before taking off on his transcontinental airplane flight, that this was his eleventh trip by aircraft, suggesting that he made six other campaign flights in the first half of 1924.

33. Ezra Meeker vs. J.A. McKinnon, "Tipsoo Lake Goal in new Road Battle," *Seattle Star*, September 6, 1924, 13.

34. "Meeker Hits Chinook Pass Road, Pioneer Finds Trip Dangerous, Tells Naches Route Advantages," *Seattle Post-Intelligencer*, September 21, 1924, 5.

35. "Meeker Runs for Legislature," *Seattle Star*, September 6, 1924, 2.

36. "Meeker Plans Fight at Good Roads Meet," *Seattle Post-Intelligencer*, December 11, 1924, 3.

37. "Meeker Makes Natches Pass Fight, Roads Body Fails to O.K. Plan," *Seattle Post-Intelligencer*, December 13, 1924, 3.

38. "Meeker Pass Plan Seen Doomed," *Seattle Post-Intelligencer*, January 28, 1925, 4.

Chapter 15: Collaboration

1. Professor of Education in English Howard Driggs went on leave from the University of Utah in 1918 to research the teaching of English in American schools. In 1923 he joined the faculty of the New York University School of Education, where he received his doctoral degree in 1926. The next year he became the chairman of the NYU English department. Driggs was well connected with literary and educational circles in New York City, where Meeker had decades of ties with the business and financial world.

2. Driggs, *Covered-Wagon Centennial*, 284–85.

3. George Himes to Ezra Meeker, November 16 and 20, 1920, Meeker Papers, Box 4, Folder 12B.

4. Howard Driggs to Ezra Meeker, January 29, 1921, Driggs Collection, B39F14.

5. World Book Company to Ezra Meeker, June 10, 1921, Driggs Collection, B39F14.

6. Eben Osborne to Ezra Meeker, June 17, 1920, Meeker Papers, Box 4, Folder 5.

7. Ezra Meeker to Howard Driggs, February 1, 1921, Driggs Collection, B39F14.

8. Howard Driggs to Ezra Meeker, March 5, 1921, Meeker Papers, Box 5, Folder 31A.

9. Howard Driggs to Ezra Meeker, March 31, 1921, Meeker Papers, Box 5, Folder 31A.

10. Howard Driggs to Ezra Meeker, April 14, 1921, Meeker Papers, Box 5, Folder 31A.

11. Howard Driggs to Ezra Meeker, May 16, 1921, Meeker Papers, Box 5, Folder 31A.

12. See Larsen, *The Missing Chapters: Ezra Meeker's Old Oregon Trail Monument Expedition January 1906 to July 1908* (Puyallup: Ezra Meeker Historical Society, 2006), 99–100 and 107–8.

13. Ezra Meeker to Howard Driggs, March 23, 1921, Driggs Collection, B39F15.

14. Howard Driggs to Ezra Meeker, March 31, 1921, Meeker Papers, Box 5, Folder 31A and World Book Company to Ezra Meeker, June 15, 1921, Meeker Papers, Box 5, Folder 31B.

15. See Dennis M. Larsen and Karen L. Johnson, *Our Faces Are Westward: The 1852 Oregon Trail Journey of Edward Jay Allen* (Independence, MO: Oregon-California Trails Association, 2012), and Johnson and Larsen, *A Yankee on Puget Sound.*

16. Ezra Meeker to Howard Driggs, May 4, 1921, Meeker Papers, Meeker Papers, Box 5, Folder 31A.

17. "Turns Back the Pages of Days on the Oregon Trail," *Pocatello Tribune,* July 28, 1921, 1.

18. George Himes to Ezra Meeker, August 6, 1921, Meeker Papers, Box 4, Folder 12B.

19. Meeker's 1921 Day Calendar, Meeker Papers, Box 11, Envelope 3, Folder 4.

20. "Borrowed Time Club Honors Pioneer," *Seattle Post-Intelligencer,* December 30, 1921, 5.

21. Peebles Publishing Company to Ezra Meeker, May 23, 1921, Meeker Papers, Box 5, Folder 31.

CHAPTER 16: MOVIE MAKING

1. Roger Blair, "The Fairbanks Medallion," *Overland Journal,* Fall 2016, 121–24.

2. "Old Oregon Trail Pioneers Guests of Baker Residents for Two Days," *Baker Morning Democrat,* July 4, 1922, 1; "Pioneer's Pageant Proclaimed as Ranking Any Other in West," *Baker Morning Democrat,* July 4, 1922, 1.

3. Walter Meacham to Avard Fairbanks, September 16, 1923, Avard Fairbanks Papers, Accn #1336, Box 11, Folder 4, Correspondence with Walter Meacham, 1923–1950, Marriott Library, University of Utah.

4. "Pioneer of Trail Arrives in Boise," *Idaho Statesman,* July 18, 1922, 5; "Ezra Meeker to Work for D.A.R.," *Idaho Statesman,* July 23, 1922, 14.

5. Jim Gentry, *In the Middle and on the Edge: The Twin Falls Region of Idaho* (Twin Falls: College of Southern Idaho, 2003), 249.

6. "The Oregon Trail," Addison E. Sheldon, *Nebraska State Journal,* September 24, 1922, 9. The Nebraska position ultimately prevailed in Congress when it designated the two routes as separate National Historic Trails.

7. "Chicagoans Given Real Thrill As Ezra Meeker Pilots His Ox-Drawn Wagon Along Michigan Boulevard," *Bellingham Herald,* October 3, 1922, 1.

8. Katherine Hunt, "Meeker Again Heeds Call of Old Oregon Trail," *Tacoma Ledger,* February 4, 1923, Magazine Section, 4.

9. Ezra Meeker to F. J. McConnell, Universal Film Mfg. Co., March 23, 1922, Meeker Papers, Box 5, Folder 32.

10. "Will Again Travel the Old Oregon Trail," *New York Times,* July 16, 1923.

11. See "The Camp Fire Girls," chapter 9.

12. Ezra Meeker to William Bonney, November 19, 1922, Meeker Papers, Box 4, Folder 7C and February 13, 1923, Meeker Papers, Box 7, Folder27B. Driggs confirms this in *Covered-Wagon Centennial,* 10.

13. Eaton (1878–1962) served in the Oregon State Legislature, opened a book and art store in Eugene, and curated the Oregon Art Room at the 1915 Panama-Pacific Exposition. In 1920 he became director of surveys and exhibits for the New York City-based Russell Sage Foundation. His work with Ezra in 1923 carried over to 1926 when he became a board member of the Oregon Trail Memorial Association.

14. Robert Bruce to Clarence Bagley, November 25, 1922, Meeker Papers, Box 7, Folder 27C.

15. Ezra Meeker to William Bonney, December 4, 1922, Meeker Papers, Box 7, Folder 27B.

16. The *Boston Post* on February 18, 1923, listed Howard R. Driggs, George W. Middleton, and Frederick W. Middleton as also being involved in the film-making project. Arthur Krouse was to be the scriptwriter.

17. Fitzgerald was a veteran of the Modoc War of 1872–73; a chapter he wrote about the various northwest Indian wars appeared in Meeker's book *Seventy Years of Progress*.

18. Ezra Meeker to William Bonney, January 3, 10, 19, 20, 22, and 31, 1923, Meeker Papers, Box 4, Folder 7C; and William Bonney to Ezra Meeker, January 17, 25, and 27, 1923, Meeker Papers, Box 4, Folder 7C.

19. Directed by James Cruze, the silent film *Covered Wagon* was released by Paramount Pictures in 1923. Meeker did not like the novel by Emerson Hough that it was based on. "I have just been reading the 'Covered Wagon'…it's the worst trash; hasn't even literary merit to commend it." Ezra Meeker to William Bonney, March 12, 1923, Meeker Papers, Box 4, Folder 7C.

20. Ezra Meeker to William Bonney, January 20, 1923, Meeker Papers, Box 4, Folder 7C.

21. Ezra Meeker to William Bonney, June 14, 1923, Meeker Papers, Box 4 Folder 7C.

22. "Radio, How the Voice Vibrates Ether," *New York Times*, July 8, 1923 and Ezra Meeker to William Bonney, June 20, 1923, Meeker Papers, Box 4, Folder 7C.

23. Ezra Meeker to William Bonney, June 20, 1923, Meeker Papers, Box 4, Folder 7C.

24. The itinerary was: Kansas City, MO, July 23; Council Bluffs, IA, July 26; Omaha, NE, July 28; Fremont, NE, July 31; Columbus, NE, August 1; Central City, NE, August 2; Grand Island, NE, August 3; Kearney, NE, August 5; North Platte, NE, August 8; Scottsbluff, NE, August 10; Casper, WY, August 13; Pocatello, ID, August 19; Boise, ID, August 26; Baker, OR, August 30; LaGrande, OR, September 2; Pendleton, OR, September 5; Walla Walla, WA, September 9; The Dalles, OR, September 16; Portland, OR, September 19. "Prospectus: Pioneers of America," Meeker Papers, Box 8, Folder 4A.

25. Ezra Meeker to William Bonney, July 1, 1923, Meeker Papers, Box 4, Folder 7C.

26. "Ezra Meeker, 92-year-old Pioneer of the Oregon Trail, is in Kansas City Again," *Kansas City Star*, July 23, 1923, Meeker Papers, Box 14, Folder 6. Photographs were taken at the mill and Meeker obtained a package of cornmeal and flour and mailed them to Bonney.

27. Robert Bruce to Clarence Bagley, August 21, 1923, Meeker Papers, Box 7, Folder 27C.

28. "Making Likeness of Pioneer," *Seattle Times*, September 11, 1923, Graphic Fiction Page.

29. "Ezra Meeker Suffers Relapse," *Seattle Times*, October 24, 1923, 12.

30. "Ezra Meeker Critically Ill," *Seattle Post-Intelligencer*, October 24, 1923, 1.

31. Ezra Meeker to Howard Driggs, December 3, 1923. Driggs Collection, B39, F15.

32. Ibid.

33. Robert Bruce to Ezra Meeker, December 7, 1923, Meeker Papers, Box 7, Folder 27C.

34. Ezra Meeker to Howard Driggs, March 5 and March 15, 1924, Driggs Collection, B39F16.

35. Howard Driggs to Ezra Meeker, March 23, 1924, Driggs Collection, B39F16.

36. Ezra Meeker to Howard Driggs, March 30, 1924, Driggs Collection, B39F16.

37. Howard Driggs to Ezra Meeker, April 6, 1924, Driggs Collection, B39F16.

38. "The City of Puyallup, with Ezra Meeker's Covered Wagon Drawn by a Yoke of Oxen Won First in the Best Ox Team Class," "Oxen, Wagons, Floats, Teams, vie in Parade," *Centralia Tribune*, August 14, 1931, 1; "Prize Won by Meeker Wagon," *Puyallup Valley Tribune*, August 14, 1931, 1.

39. Andy Anderson, email correspondence, November 3, 2014.

Chapter 17: The World's Oldest and Youngest Aeronaut

1. "Ezra Meeker Has a New Experience," *Seattle Post-Intelligencer*, September 26, 1919, 4.

2. A search of all likely depositories in Victoria has failed to turn up this manuscript, which may have told of his 1858 journey to Victoria. There is a listing at the end of *Kate Mulhall* of an 1874 publication, which seems to also be missing. Perhaps it is the text Meeker brought to Victoria in 1920.

3. "Aged Pioneer Makes Trip by Air," *Seattle Post-Intelligencer*, May 27, 1921, 16.

4. Kelly continued to make flying records. In 1925 he set the round trip speed record from San Francisco to Vancouver at ten hours and fifty-five minutes.

5. In this Meeker succeeded. Airmail service to Pasco began on April 6, 1926.

6. "Meeker Races Storm in Trail Blazing Flight," *Seattle Post-Intelligencer*, October 2, 1924, 1.

7. Ibid.

8. "Meeker Plane Hits Pocket, Wakens Ezra," *Seattle Post-Intelligencer*, October 3, 1924, 1.

9. "Ezra, Air Flight Over, On Lookout for New Thrills," *Seattle Post-Intelligencer*, October 3, 1924, 1.

10. "Meeker in Illinois Gets Thrill from Wind Storm Aloft," *Seattle Post-Intelligencer*, October 4, 1924.

Notes for pages 188–195 249

11. "Meeker Reaches Rantoul," *Tacoma Daily Ledger*, October 4, 1924, 1.

12. "Ezra on a Flying Trip," *Omaha World Herald*, October 4, 1924, 1.

13. Ibid.

14. "Ezra Air Flight Over, On the Lookout for New Thrills," *Seattle Post-Intelligencer*, October 5, 1924, Section HH, 1.

15. The photograph of Meeker meeting President Coolidge on the White House lawn took place on this occasion, not at the time of the signing of the 1926 Oregon Trail commemorative coin bill as some allege. Driggs, *Covered-Wagon Centennial*, 296.

16. "Meeker Stars on Program," *Seattle Post-Intelligencer*, January 18, 1925, 9.

CHAPTER 18: "NO HALF HOURS TO SPARE"

1. Michael Wallis, *The Real Wild West: The 101 Ranch and the Creation of the American West* (New York: St. Martin's Press, 1999), 493, 497, 608, and 642.

2. "Ezra Meeker in Wild West Show," *Tacoma News Tribune*, February 19, 1925, 1.

3. "Meeker to Hit Circus Sawdust Trail," *Tacoma News Tribune*, March 17, 1925, 9. "Ezra Meeker Runs Away, Joins Circus," *Seattle Post-Intelligencer*, March 17, 1925, 10.

4. Gary M. Greenbaum, "Sunrise at Pocatello: Dr. Minnie Howard, Ezra Meeker and the Oregon Trail Half Dollar," *The Numismatist*, November 2014, 35–40.

5. Clara B. Dice Roe, *The Prairie Years* (Philadelphia: Dorrance & Company, 1957), 21–31.

6. Robert Bruce to William Bonney April 26, 1926, Meeker Papers, Box 5 Folder 32.

7. Robert Bruce to Clarence Bagley, October 5, 1926, Meeker Papers, Box 7, Folder 27C.

8. Bagley also predicted that only one half million Oregon Trail commemorative coins would be sold for the same reason.

9. Robert Bruce to Clarence Bagley, October 11, 1926, Meeker Papers, Box 7, Folder 27C.

10. George Himes to William Bonney, October 9, 1926, Meeker Papers, Box 4, Folder 12B.

11. William Bonney to George Himes, October 12, 1926. See Larsen, *Hop King*, 188–90 for the bank story.

CHAPTER 19: THE OREGON TRAIL MEMORIAL ASSOCIATION

1. When the dams were built water levels fell short of flooding the site. A simple stone marker stands today at the location about twelve miles west of Fort Hall, Idaho. The irony is that a simple miscalculation of water levels by Minnie Howard and her Idaho supporters eventually led to the culmination of many of Meeker's dreams.

2. The Stone Mountain, Georgia, fifty-cent piece was struck in 1925. Its purpose was to raise funds for the Confederate generals monument that was being carved into the face of Stone Mountain.

3. Gary M. Greenbaum, "Sunrise at Pocatello: Dr. Minnie Howard, Ezra Meeker and the Oregon Trail Half Dollar," *The Numismatist*, November 2014, 35–40.

4. Driggs, *Covered-Wagon Centennial*, 323–24.

5. Driggs, *Covered-Wagon Centennial*, 16. The official goals of the Oregon Trail Memorial Association were 1) To identify and mark the line of the Oregon Trail; 2) To erect suitable monuments at important sites and landmarks along the trail; 3) To restore the Whitman Mission and promote the establishment of a park at the site and to identify and mark the site of other massacres along the trail; 4) To promote in moving pictures the story of the trail for the purpose of teaching exact and truthful history to the schools of the nation; 5) To collect and preserve written accounts of the trail. "What Has Already Been Accomplished," Bulletin No. 4 (New York: OTMA 1926), Meeker Papers, Box 8, Folder 2.

6. Congressman Addison Smith to Clarence Bagley, December 10, 1925, Meeker Papers, Box 7, Folder 27B.

7. Meeker supplied the date in an appendix to his novel *Kate Mulhall: A Romance of the Oregon Trail* (New York: Ezra Meeker and Oregon Trail Memorial Association, 1926), 285.

8. The officers were: Ezra Meeker, president; Charles W. Davis and Chauncey Depew, vice presidents; David G. Wylie, secretary with Robert Bruce as his assistant; and Martin S. Garretson, treasurer. The Executive Committee consisted of Meeker, Seymour, Allen Eaton, Palmer Rogers, and Guthrie Y. Barber.

9. Driggs, *Covered-Wagon Centennial*, 11.

10. Ezra Meeker to William Bonney, February 6, 1926, Meeker Papers, Box 7, Folder 27A.

11. Greenbaum, "Sunrise at Pocatello," 34.

12. Driggs, *Covered-Wagon Centennial*, 14, 324.

13. Robert Bruce to William Bonney, March 4, 1926, Meeker Papers, Box 5, Folder 32.

14. Robert Bruce to Allen Eaton, April 19, 1926, Meeker Papers, Box 7, Folder 27C.

15. Ibid.

16. Robert Bruce to Clarence Bagley, April 16, 1926, Meeker Papers, Box 7, Folder 27C.

17. Greenbaum, "Sunrise at Pocatello," 35.

18. Greenbaum, "Sunrise at Pocatello," 36.

19. Q. David Bowers, *Commemorative Coins of the United States: A Complete Encyclopedia* (Wolfeboro, NH: Bowers & Merena Galleries, Inc., 1992), 208. Carl Stang, "Canine and Equine: The Art of Laura Gardin Fraser," July 2013, *The Numismatist*, 35.

20. Lawrence L. Dodd. "Review of Bert Webber, The Oregon Trail Memorial Half-Dollar," *Overland Journal*, Spring 1987, 48–49.

21. Maue made the entire distance from New York to Portland, Oregon, with Meeker. There is no record of discord between the two men as occurred in the 1916 Pathfinder expedition between Meeker and several of his drivers.

22. Meeker argued that on the day of his first birthday he was starting his second year of life. So on the day of his ninety-fifth birthday he was starting his ninety-sixth year of life. "Silver Paves Old Trail; Ezra Hits Road Again," *Seattle Post Intelligencer*, July 30, 1926, 2.

23. "Meeker to Motor Over the Oregon Trail," *New York Times*, July 11, 1926, 24.

24. "Meeker Car Making Time," *New York Times*, July 25, 1926, 7.

25. "$100 Real Money When Ezra Was Boy; Now Refuses to Run Race for Less," *Seattle Post-Intelligencer*, July 27, 1926, 1.

26. "Oregon Trails Days Fiesta Is an Outstanding Success," *Gering* [Nebraska] *Courier*, July 30, 1926, 1.

27. "Ezra Meeker Pioneer of Oregon Trail, Continues Journey West; To Make the Return Trip Soon," *Scottsbluff Star-Herald*, July 31, 1926, 1.

28. "Trail Blazer Revisits Old Spring," *Oregonian*, August 13, 1926, 4.

29. The Philadelphia Mint struck 48,000 coins in September. They sold rapidly and, expecting the bounty to continue, OTMA requested an additional 100,000 coins to be struck in San Francisco. The second batch failed to sell out and the Treasury ordered a halt to further minting until all the 1926 coins were purchased.

30. "Trail Blazer Revisits Old Spring," *Oregonian*, August 13, 1926, 4.

31. "Trail-To-Rail-Celebrated," *Seattle Post-Intelligencer*, August 20, 1926, 5.

32. The statue, by sculptor Alonzo Victor Lewis, stands in Pioneer Park in Puyallup next to an ivy-covered pergola that marks the site of the cabin in which the family lived from 1862 to 1889 until they moved into the Meeker Mansion.

33. "Reverent Crowd Witnessed the Unveiling Ceremonies When Ezra M. Meeker Was Highly Honored," *Puyallup Valley Tribune*, September 18, 1926, 1.

34. "Ezra Meeker at Work on Oregon Trail Program," *Puyallup Valley Tribune*, September 11, 1929, 1.

35. "Meeker Urges Oregon Trail Coinage Use," *Seattle Post-Intelligencer*, September 9, 1926, 17.

36. "Spokane to Hear Old Trail Blazer," *Seattle Post-Intelligencer*, October 5, 1926, 13.

37. "Ezra Meeker Parades Today," *Seattle-Post Intelligencer*, November 15, 1926, 12.

38. "What's a Trifle Like Rain to Ezra Meeker?" *Seattle Post-Intelligencer*, November 16, 1926, 16.

39. Driggs Collection, Box 38, Folder 75.

40. Robert Bruce to Clarence Bagley, September 29, 1926, Meeker Papers, Box 7, Folder 27C.

41. Clarence Bagley to Robert Bruce, October 6, 1926, Meeker Papers, Box 7, Folder 27C.

42. Robert Bruce to Clarence Bagley, October 5, 1926, Meeker Papers, Box 7, Folder 27C.

43. Robert Bruce to William Bonney Bagley, October 20, 1926, Meeker Papers, Box 7, Folder 27C.

44. Robert Bruce to William Bonney, November 1, 1926, Meeker Papers, Box 7, Folder 27B.

45. Robert Bruce to Clarence Bagley, December 9 and 16, 1926, Meeker Papers, Box 7, Folder 27C.

46. Clarence Bagley collection, N979.719, University of Washington Library.

47. A copy of the agreement may be found in the Meeker Papers, Box 7, Folder 27C.

48. Edmund Seymour to William Bonney, August 23, 1927, Meeker Papers, Box 7, Folder 27B.

49. Greenbaum, "Sunrise at Pocatello," 37.

50. Ezra Meeker to William Bonney, December 21, 1927, Meeker Papers, Box 4, Folder 7C.

51. Many coin historians accepted and repeated the three million dollar income figure for OTMA's coin sale, but simple math shows this cannot be correct. The number of coins actually sold was 203,006. Dividing that number by two, as fifty cents had to be paid to the U.S. treasury for each coin, left $101,503. This sum was divided among the OTMA, the Equitable Trust Company (and later the Scott Coin and Stamp Company), and the local banks that were selling the coins. It also had to cover shipping costs. The only way to arrive at three million dollars is to assume that six million coins were minted and sold and that the OTMA realized all the income beyond the fifty-cent purchase price. This did not happen. Bruce J. Noble Jr., "Marking Wyoming's Oregon Trail," *Overland Journal* (Summer 1986), 21.

52. "What Happened to the Profits," in Bowers, *Commemorative Coins of the United States*. Bowers, who is considered in coin circles the authority on commemorative coins, made other errors in his telling. He stated that Meeker went over the Oregon Trail in 1851, and that the coin was introduced on the 75th anniversary of that crossing. Meeker traveled the Oregon Trail in 1852. Bowers, 213.

53. Bowers began the section on the coin in his book quoting historian Arlie Slabaugh, who claimed that while, artistically, the coin was his favorite, "from an ethical standpoint it is not." He then proceeded to cast doubt on the ethics of the OTMA leaders throughout the remainder of his discussion of the coin. Bowers, 207.

54. The same argument is made by Will Rossman in, "Ezra Meeker and the Oregon Trail Commemorative," *The Numismatist*, August 1998, 887–90. Perhaps the fact that Congress said in the Act that only the OTMA could purchase the coins, thus giving the OTMA a monopoly, upset coin collectors who could not order from the mint.

55. Gary M. Greenbaum, "The Other Side of the Oregon Trail Half Dollar: The Oregon Trail Memorial Association's View of Its Commemorative," *The Numismatist* (October 2013), 43–49.

56. Bert Webber, *The Oregon Trail Memorial Half-Dollar, 1926–1929* (Medford, Oregon, Webb Research Group, 1986), 44.

Notes for pages 211–216   253

57. Thomas B. Hill to William Bonney, May 18, 1931, Meeker Papers, Box 7A, Folder 27C.

58. William Bonney to Mrs. Eliza F. Leary, Seattle, and W. L. McCormick, Tacoma, November 20, 1931, Meeker Papers, Box 7A, Folder 27C.

CHAPTER 20: END OF THE TRAIL

1. Ezra Meeker to William Bonney, May 11, 1926, Meeker Papers, Box 4, Folder 7C.

2. Ezra Meeker to OTMA membership, July 7, 1928, Meeker Papers, Box 14.

3. Ibid.

4. Telegram, Ezra Meeker to William Bonney, June 22, 1928, Meeker Papers, Box 4, Folder 7C.

5. Driggs, *Covered-Wagon Centennial*, 300.

6. The five previous trips were the 1906 monument expedition, the 1910 mapping expedition, the 1916 Pathfinder automobile expedition, the 1923 Pioneers of America movie venture, and the 1926 OTMA memorial coin trip. Left out of this accounting was his 1922 trip, the 1924 airplane flight over the route of the trail, and his innumerable trips from Puyallup and Seattle to New York by rail during his hop growing years, and multiple trips to various individual Oregon Trail sites such as Fort Hall.

7. "Ezra Meeker to Hit Oregon Trail Again," *New York Times*, August 4, 1928, 15.

8. The itemized bill can be found in the Driggs collection, B39 F18.

9. What happened to the Oxmobile? Andy Anderson, President of the Puyallup Historical Society at the Meeker Mansion, made a visit to the Ford Museum in July 2007 and learned that the Oxmobile was put in storage in Dearborn, but not accessioned into the permanent collection. It was occasionally refurbished for use in the OTMA's celebrations. The last recorded request by the society was in 1943. There is no trace of the Oxmobile today.

10. "Ezra Meeker Home Again," *Seattle Post Intelligencer*, November 1, 1928, 9.

11. The Frye hotel was built by George Frye and his wife Louise Denny Frye, daughter of Arthur Denny, one of the founders of Seattle. In 1853 Arthur Denny was canoeing home from a gathering of the territorial legislature at Olympia and stopped overnight at Meeker's cabin on McNeil Island. In a stroke of irony his daughter would supply Meeker with his final dwelling place.

12. This was likely the last photograph of the multiple hundreds taken of Meeker throughout his life.

13. "My Work Is Not Finished," Lori Price Papers, The Puyallup Historical Society at the Meeker Mansion.

14. "Children to Visit Bier of Ezra Meeker," *Seattle Post-Intelligencer*, December 4, 1928, 1.

15. William Bonney located the wagon in Portland, stored in a Ford Motor Company facility. The oxen came from the Valley Gem Farms in Arlington and were trucked to Seattle. "Pioneer's Old Wagon Will Bring Tribute," *Seattle Post-Intelligencer*, December 4, 1928, 3 and "Oxen, Wagon, Tribute to Ezra Meeker," *Seattle Post-Intelligencer*, December 5, 1928, 1.

16. "Cortege Takes Ezra Meeker to End of Life's Long Trail," *Seattle Post-Intelligencer*, December 6, 1928, 3.

Afterword

1. Larsen, *Hop King*, 127, 225n2. The complete story of the Ballards and Meeker may be found in the spring 2012 issue of *Northwest Trails Journal* at www.octa-trails.org/chapters/north-west/newsletter.php; Roy Page Ballard, "Story of his Family," 1945, ballardfamily.wordpress.com/about.

2. Ezra Meeker to Eliza Egbert, May 7, 1903, Meeker Papers, Box 6, Vol. 2.

3. Dennis M. Larsen, *Slick as a Mitten: Ezra Meeker's Klondike Enterprise* (Pullman: Washington State University Press, 2009), 37.

4. See Larsen, *Hop King*, 6–7, for the story of the adoption.

Bibliography

Archival Sources

Southern Utah University, Gerald R. Sherratt Library, Special Collections, Cedar City, UT. Holds the Howard R. Driggs Collection. Box 23, Folder 11, contains the 1920 contract between Meeker and Driggs regarding *Ox-Team Days on the Oregon Trail*. Box 39 contains items and stories relating to Meeker and correspondence between the two men. It also contains a number of photographs. www.li.suu.edu/page/special-digital-collections-howard-r-driggs-collection-about.

University of Utah, Marriott Library, Salt Lake City, UT. Avard Fairbanks Papers. Box 11, Folder 4, correspondence with Walter Meacham, 1923–1950.

Oregon-California Trails Association Mattes Library, Independence, MO. Holds a copy of Jacob Resser's unpublished 1852 Oregon Trail Diary along with a transcription.

Oregon State Historical Society, Portland, Oregon. Holds material about Meeker and George Himes, including photographs and correspondence.

Puyallup Historical Society at the Meeker Mansion, Puyallup, WA. Holds the Lori Price papers, scrapbooks, and various items pertaining to Meeker's life and thirty-five binders of photocopies of Meeker's correspondence.

University of Washington Libraries, Special Collections, Seattle, WA. Holds Clarence Bagley papers, 1864–1931; Ezra Meeker materials and photographs; and the Edmond S. Meany papers, 1877–1935, including Meany's "Living Pioneers of Washington" scrapbook, 1915–1920.

Washington State Historical Society Research Center, Tacoma, WA. Holds the Ezra Meeker Papers, seventeen archival boxes that may be viewed by appointment. A large binder contains photocopies of the photographs, and some photographs may be viewed online. A general index of the documents and letters, created by Frank Green in 1969, is available at various libraries and at the research center. The author intends to donate his photocopies of Meeker's correspondence to the Puyallup Historical Society at the Meeker Mansion upon the publication of this work where they will be more readily available to the public. Two items of special interest are the prospectus for Meeker's 1923 Pioneers of America Organization which is located in Box 8 Folder 4A and the incorporation document for the OTMA found in Box 8 Folder 2.

Personal Correspondence

Andy Anderson, historian, Puyallup Historical Society at the Meeker Mansion, Puyallup, WA.

Patrick Harris, director, Aurora Colony Museum, Aurora, OR.

Nancy Jackson, Washington State Historical Society Research Center, email, May 2, 2017.

Web Resources

Ballard Family. ballardfamily.wordpress.com/about.
Biographical Directory of the United States Congress, 1774-Present. bioguide.con gress.gov/biosearch/biosearch.asp.
Congressional Delegations from Washington. HistoryLink.org. www.historylink.org/File/5463.
Inflation Calculator. www.in2013dollars.com/1890-dollars-in-2015?amount=30268.
United States Census (various years). www.digitalarchives.wa.gov.
William Mardon. www.findagrave.com/memorial/112877129/william-bruce-mardon.

Manuscripts, Books, Articles

Abbott, Carl. *The Great Extravaganza: Portland and the Lewis and Clark Exposition.* Portland: Oregon Historical Society Press, 1996.
Bagley, Clarence. *History of Seattle: From the Earliest Settlement to the Present Time.* Vol. 3. Chicago: S. J. Clarke Publishing Co., 1916.
Blair, Roger. "The Fairbanks Medallion." *Overland Journal,* Fall 2016.
———. "I Remember the Day: A Connection to Ezra Meeker and Oregon Trail History," interview with Effie Ritchey, *News from the Plains,* vol. 12, no. 1, January 1998, Oregon-California Trails Association.
Bonney, William P. *History of Pierce County, Washington.* Chicago: Pioneer Historical Publishing Company, 1927.
Bowers, David Q. *Commemorative Coins of the United States: A Complete Encyclopedia.* Wolfeboro, NH: Bowers & Merena Galleries, Inc., 1991.
Canse, John M. *Pilgrim and Pioneer Dawn in the Northwest.* New York: The Abingdon Press, 1930.
Chapman, Arthur. "The Covered Wagon Centennial," *New York Herald Tribune,* April 6, 1930.
Corell, Harris Anderson. "A Unique Return Trip over the Famous Oregon Trail." *Overland Journal,* June 1910.
Driggs, Howard. Ed. *Covered-Wagon Centennial and Ox-Team Days.* Yonkers-on-Hudson, New York: World Book Company, 1932.
Eichhorst, Jerry. "Building Fort Hall: A Story Told By Many People." *Overland Journal,* Spring 2016.
Egan, Ray. "On the Trail of a Legend: The Ox Rope Story." *Overland Journal,* Summer 2016.
Gentry, Jim. *In the Middle and on the Edge: The Twin Falls Region of Idaho.* Twin Falls: College of Southern Idaho, 2003.
Green, Frank. *Ezra Meeker, Pioneer: A Bibliographical Guide.* Tacoma: Washington State Historical Society, 1969.
———. "The Three Musketeers of Northwest History: Early Chroniclers Meeker, Bagley and Himes Built a Historical Foundation." *Columbia,* Winter 1989.
Greenbaum, Gary M. "The Other Side of the Oregon Trail Half Dollar: The Oregon Trail Memorial Association's View of its Commemorative." *The Numismatist,* October 2013.

———. "Sunrise at Pocatello: Dr. Minnie Howard, Ezra Meeker and the Oregon Trail Half Dollar." *The Numismatist*, November 2014.
Holmes, Kenneth L. and David C. Duniway, eds. "Eliza Ann McAuley, Iowa to the Land of Gold." *Covered Wagon Women: Diaries and Letters from the Western Trails, 1852*, Vol. 4. Glendale: Arthur H. Clark, 1986.
Hunt, Herbert. *Tacoma, Its History and Its Builders: A Half Century of Activity*, Vol. 1 (Chicago: S. J. Clarke Publishing Company, 1916), 128 and 137.
Johnson, Karen L. and Dennis M. Larsen. *A Yankee on Puget Sound: Pioneer Dispatches of Edward Jay Allen, 1852–1855*. Pullman: WSU Press, 2013.
Klugar, Richard. *The Bitter Waters of Medicine Creek: A Tragic Clash Between White and Native Americans*. New York, Vintage Books, 2011.
Larsen, Dennis M. "The Ballard Family on the Oregon Trail in 1852." *Northwest Trails* 28, no. 1 (Spring 2013). www.octatrails.org/chapters/northwest/newsletter.php.
———. *Ezra Meeker: A Brief Resume of His Life and Adventures*. 2nd ed. Puyallup: Puyallup Historical Society at Meeker Mansion, 2013.
———. "Ezra Meeker's March to Washington: March–December 1907." *Overland Journal*, Summer 2006.
———. "A Fatal Intersection." *Tumwater Historical Association Newsletter*, Vol. 35, Issue 3 (Winter 2016).
———. *Hop King: Ezra Meeker's Boom Years*. Pullman: WSU Press, 2016.
———. *The Missing Chapters: The Untold Story of Ezra Meeker's Old Oregon Trail Monument Expedition January 1906 to July 1908*. Puyallup: Ezra Meeker Historical Society, 2006.
———. *Slick as a Mitten: Ezra Meeker's Klondike Enterprise*. Pullman: WSU Press, 2009.
Larsen, Dennis M. and Karen Johnson. *Our Faces Are Westward: The 1852 Oregon Trail Journey of Edward Jay Allen*. Independence, MO: Oregon-California Trails Association, 2012.
Lockley, Fred. "Ezra Meeker the Great Spirit of the West." *Overland Monthly*, February 1923.
———. "The Old Immigrant Trail: The Story of Ezra Meeker and His Ox-Team," *Pacific Monthly*, Vol. 17, January 1907.
Magnusson, Elva Cooper. "The Naches Pass." *Washington Historical Quarterly* 25, July 1934.
Mattes, Merrill J. "A Tribute to the Oregon Trail Memorial Association." *Overland Journal*, Winter 1994.
Meany, Edmund. *Living Pioneers of Washington*. Seattle Genealogical Society, 1995.
Meeker, Ezra. *The Busy Life of Eighty-Five Years of Ezra Meeker*. Indianapolis: Wm. B. Burford Press, 1916.
———. *Hop Culture in the United States*. Puyallup: E. Meeker and Co., 1883.
———. *Kate Mulhall: A Romance of the Oregon Trail*. New York City: Ezra Meeker and Oregon Trail Memorial Association, 1926.
———. *The Ox Team or the Old Oregon Trail, 1852–1906*. 1st ed. Lincoln NE: Jacob North & Co., 1906. 2nd and 3rd eds. Chicago: W. A. Donahue & Co., 1906 and 1907. 4th ed. Syracuse, New York: Mason Press Co., 1907.
———. *Personal Experiences on the Old Oregon Trail Sixty Years Ago*. St. Louis: McAdoo and Company, 1912.

———. *Pioneer Reminiscences of Puget Sound: The Tragedy of Leschi*. Seattle: Loman & Hanford, 1905.
———. *Seventy Years of Progress in Washington*. Tacoma: Allstrum Printing Company, 1921.
———. "Speech at Old Fort Hall." *Overland Journal*. Independence, MO: Oregon-California Trails Association, Spring 2016.
———. *Story of the Lost Trail to Oregon*. Seattle: 1910. Rpt., Fairfield, Washington: Ye Galleon Press, 1998.
———. *Story of the Lost Trail to Oregon No. 2*. Seattle, 1917.
———. *Uncle Ezra's Short Stories for Children*. Tacoma: 1912.
———. *Ventures and Adventures of Ezra Meeker*. Seattle: Rainier Printing Co., 1909.
———. *Washington Territory West of the Cascade Mountains*. Olympia: Transcript Office, 1870.
Meeker, Ezra and Howard Driggs. *Ox-Team Days on the Oregon Trail*. Yonkers-on-Hudson, New York: World Book Company, 1925.
Noble, Bruce J. Jr. "Marking Wyoming's Oregon Trail." *Overland Journal*, Summer 1986.
Price, Lori and Ruth Anderson. *Puyallup: A Pioneer Paradise*. Charleston: Arcadia, 2002.
"Reception to Ezra Meeker on the Evening of February 20, 1908." *Washington State Historical Society Publications*, Vol. 2, 1907–1914. Olympia: Frank M. Lamborn, Public Printer, 1915.
Richards, Kent D. *Isaac I. Stevens Young Man in a Hurry*. Pullman: Washington State University Press, 1993.
Roe, Clara B. Dice. *The Prairie Years*. Philadelphia: Dorrance & Company, 1957.
Rossman, Will. "Ezra Meeker and the Oregon Trail Commemorative." *The Numismatist*, August 1998.
Roth, Richard R. *The Hot Lake Story: An Illustrated History from Pre-discovery to 1974*. Orting, WA: Heritage Quest Press, 2008, and *The Hot Lake Supplement*, 2009.
Sheldon, Addison E. "The Oregon Trail." *Nebraska State Journal*, September 24, 1922.
Stang, Carl. "Canine and Equine: The Art of Laura Gardin Fraser." *The Numismatist*, July 2013.
Taylor, Quintard. "Slaves and Free Men Blacks in the Oregon Country, 1840–1860." *Oregon Historical Quarterly*, Summer 1982.
Umatilla County Historical Society, "The Day Ezra Meeker Came to Our School," *Pioneer Trails*, Summer 1990, 14–16.
Wallis, Michael. *The Real Wild West: The 101 Ranch and the Creation of the American West*. New York: St. Martin's Press, 1999.
Webber, Bert. *Ezra Meeker: Champion of the Oregon Trail*. Medford, OR: Webb Research Group, 1998.
———. *The Oregon Trail Memorial Half-Dollar (1926–1939)*. Medford, OR: Webb Research Group, 1986.

Newspapers and Other Publications

The Washington State Library in Tumwater, Washington, has most of the state newspapers on microfilm and several from out of state. The collection may be accessed by appointment only. The Northwest Room at the Tacoma Public Library and the University of Washington Library also maintain fairly extensive collections

Bibliography   259

of state newspapers on microfilm. Many other repositories of newspapers are online, some for a fee, some free. Below are newspapers I found useful.

WASHINGTON STATE

Aberdeen World
Arlington Times
Bellingham Herald
Bremerton Searchlight
Centralia Chronicle
Chehalis Bee Nugget
Cowlitz County Advocate
Ellensburg Capital
Enumclaw Herald
Everett Herald
Morning Olympia

Puyallup Tribune
Puyallup Valley Tribune
Seattle Post-Intelligencer
Seattle Star
Seattle Times
Tacoma Daily Ledger
Tacoma News
Waitsburg Times
Washington Standard (Olympia)
Weekly Pacific Tribune (Olympia)
Yakima Daily Republic

OTHER NEWSPAPERS AND MAGAZINES

Baker Morning Democrat (Baker City, OR)
Bridgeport (CT) Herald
Bridgeport (CT) Times-Star
Casper Record
Daily Capital Journal (Salem, OR)
Eugene Register Guard
Gering (NE) Courier
Idaho Statesman (Boise)
Illinois State Register (Springfield)
Kansas City Star
Long Beach Daily Telegram
Milton-Freewater (OR) Valley Herald
New York Times

Omaha World Herald
Oregonian (Portland)
Pacific Monthly (Portland, OR)
Pocatello Tribune
Redlands Daily Facts
San Bernardino Sun
San Diego Union
San Francisco Call
Scottsbluff Star-Herald
Topeka Capital Journal
Wind River Mountaineer (Lander, WY)
Wyoming State Journal (Laramie)

Index

101 Ranch Wild West Show, 190; Bison Film Company, 191

Ackerson, John: named Tacoma, 13
Ackerson, Louise, 13, 139, 226n19;
AYPE loan, 44, 72, 84
Alaska-Yukon-Pacific-Exposition: warning 39; exhibit, 41; disappointing start, 44
Albrecht, Oscar A., 98; filming Meeker, 100
Albro, Joseph W., 157, 243n10
Allen, Edward Jay, 35, 76, 77, 153, 165, 211; built 1853 road, 156; Oregon Trail Manuscript, 168
Allen, Harold, 168, 189
Allen, James, 162
Anderson, William, 156
Anthony, Susan B., 29

Bagley, Clarence, 16, 198, 215, 216; 1919 trip over Naches Pass, 155; *Kate Mulhall*, 193; on coin design, 206
Ballard, Ling, 55
Barber, Guthrie Y., 197, 214
Barnes, Percy, 184
Barrett, Amos, 24
Barrett, Judson, 96
Bashford, Herbert, 51
Bashor, George, 102
Bean, Frances (Meeker's niece), 41
Bean, Vida, 41
Berger Brothers, 43
Blalock, Dr. Nelson, 59
Blethen, Alden, 59
Bloom, Murray, 76
Bogart, Mrs. N. M., 21
Bonney, William, 117, 182, 198, 216; 1919 trip over Naches Pass, 155;

Naches Pass lobby effort, 175; OTMA board, 209; Yale OTMA monument, 211
Books: *Kate Mulhall A Romance of the Oregon Trail* (1926), 193; *Ox–team Days on the Oregon Trail* (1922), 168; *Personal Experiences on the Oregon Trail* (1912), 93; *Pioneer Reminiscences and the Tragedy of Leschi* (1905), 9; *Seventy Years of Progress* (1921), 167; *Story of the Lost Trail to Oregon, No. 1*, 121; *Story of the Lost Trail to Oregon, No. 2* (1916), 129; *The Busy Life of Ezra Meeker's Eighty-Five Years* (1916), 121; *The Ox Team* (2nd and 3rd edition), 75; *Uncle Ezra's Short Stories For Children* (1912), 105; *Ventures and Adventures of Ezra Meeker* (1909), 41, 75
Borrowed Time Club, 170
Bowers, Q. David, 210
Breede, Adam, 72
Brougham, Royal, 189
Brown, Cora Osborne (Meeker's granddaughter), 80
Bruce, Robert, 33, 174, 176, 180, 193, 198; complaints about Meeker, 206; on coin design, 206
Buckwalter, Mr., 101
Buskirk, F. G., 121

Calder, Willis, 49, 50, 55, 65, 67, 68, 69, 72; loan, 70; printed *Ventures and Adventures*, 46
Camp Fire Girls, 111-112, 237n22; See also Pratt, George
Cannon, Joe, 34
Cantwell, Professor James W., 90
Carlyon Pass (White Pass), 158

Carlyon, Senator Phillip H., 158
Chambers, George, 157
Chicago parade, 172
Chicago World's Fair: Meany conflict, 17
Chittenden, Hiram, 60
Churchill, Major L. S., 187
Cline, George, 227n8
Connelly, William, 127
Coolidge, President Calvin, 198, 199, 206
Cooper, Tex, See also 101 Ranch
Corey, Arthur Robert, 87
Covered wagon: 1906 version, 35; 1923 version, 176, 182
Cowlitz Trail, 26, 240n45
Crandall, Mary, 138
Crane, Richard E., 36
Curt Teich & Company, 76, 84, 90, 115

Daughters of the American Revolution (DAR), 64, 71, 75, 93, 172, 185, 240n45; financed book, 94; role in coin project, 191
Davis, Charles, 175; National Highways Association, 134, 174; OTMA, 197
Davis, Dwight F., 186
Davis, Henry Clay, 27
de Long, Uncle Henry, 202
Denny, Arthur, 253n11
Depew, Chauncey M., 197
Dice, Clara Meeker, 191
Dog Jim, 29, 52, 103
Douglas, Walter B., 126
Driggs, Howard, 23, 166, 179, 181, 199, 206, 213, 245n1; 1923 film company, 175; coin idea, 197; wrote *Ox Team Days*, 167
Du Bois, Claude W., 109

Eaton, Allen, 175, 180, 181, 247n13; film partner, 174
Eckler, J .P., 21
Eddyville, IA, 29, 87
Edison, Thomas A., 198

Egbert, Eliza McAuley, 10
Equitable Trust Company, 252n51: coin arrangement, 207

Fairbanks, Avard, 172, 179
Farrell, Dillon, 21, 227n14
Federated Women's Club, 51
Fenton, William D., 20
First boulder marked, 231n13
Fitzgerald, Maurice, 176, 180, 216, 247n18; involvement in OTMA take over, 209
Fletcher, Charles, 128
Ford, Henry, 120, 212, 214
Fort Hall, Idaho, 130; 1921 Meeker speech, 169; monument, 195; search for, 70, 129
Fraser, James Earl, 201
Fraser, Laura Gardin, 201
Frey Hotel, 253n11
Fuller, Edward N., 8, 9; Meeker clash, 8

Garretson, Martin S., 197
Gear, Eva, 12
Gentry, Elizabeth Butler, 75, 93, 94
Gibbs, George, 17
Giesy, Rudolph, 35
Gilstrap, William, 154, 155
Gobel, Herman, 19, 27
Gobel, Juanette, 27
Gobel, Tillie, 19, 27, 226n7
Goode, Henry W., 20
Gooding, Senator Frank, 199
Governors: Albert Rosellini (Washington), 165; Bryant Brooks (Wyoming), 128; Ernest Lister (Washington), 116; Isaac Stevens (Washington), 12, 17; James Brady (Idaho), 68; John McGraw (Washington), 38, 39, 40; Louis F. Hart (Washington), 161, 162, 164; Moses Alexander (Idaho), 130, 132
Graham, Katherine, 12, 30, 31, 47, 80, 229n26
Greeley, Horace, 111

262 *Saving the Oregon Trail*

Green, Howard, 157
Greenbaum, Gary, 195
Grimes, Joe, 88, 91, 99
Guggenheim trophy automobile race, 119

Halley comet, 68
Harding, President Warren, 172
Harlan, Edgar R., 87
Harriman, Mrs. Edward, 198
Hartman, John Jr. (Meeker's attorney), 13, 85
Hayes, William, 27
Hebbard, Charles, 151
Hewitt, Henry, 58, 85
Hill, Jennie Lester, 43, 45
Hill, Samuel, 61
Himes, George, 19, 26, 60, 154, 166, 170, 182, 194, 216; 1910 and 1919 Naches Pass trips, 155
Hodgson, Caspar W., 112, 166, 197; introduced Meeker to Pratt, 173
Hood, Charles, 43, 52, 53, 103
Hoover, Herbert, 149, 150, 151, 242n5
Hot Lake Sanitarium, 64, 70
Howard, Dr. Minnie, 129, 130, 169, 199, 201, 249n1; coin, 199, 206, 209; Fort Hall monument, 195
Howard, Dr. William F., 130
Hubbard, Eddie, 184
Huggins, Edward, 10
Hughes, Charles, 135
Humphrey Bill, 34, 57
Humphrey, Congressman William, 34, 54, 76, 78, 83, 93, 104, 125, 196; introduced first Meeker bill, 57

Irwin, Charles Burton, 99

Johnson, Claire Skinner, 191
Johnson, Karen, 169
Jones, Meryl, 151
Jones, Richard Seley, 115
Jones, Senator Wesley, 104, 107, 149, 160, 198, 236n42
Joy Wheel (AYPE carnival ride), 45, 46
Kalley, Ellen Hartwig, 86

Kate Mulhall: Meeker's purpose in writing, 193
Kelly, Lieutenant Oakley, 186
Kinney, Colonel C. C., 187

Lane, William, 155
Leschi, Chief: 2nd trial acquittal vote, 21; judicially murdered, 10; Medicine Creek Treaty, 16; portrait story, 10, 225n9
Lewis and Clark Exposition, 18
Lewis, A. W., 39
Lewis, Alonzo Victor, 251n32
Lincoln Highway competition, 137
Loans: dispute over Ackerson loan, 84; requested Eben Osborne, 62, 103; requested Jacob Meeker, 92; requested Joe Templeton and Carrie Osborne, 44; requested John Hartman, Jr., 85; requested Louise Ackerson, 13, 44; requested Metropolitan Press, 13; requested Robert Wilson, 11; requested William Bonney, 123
Longmire, David, 155, 156, 160, 161
Lord, Elizabeth Laughlin, 26, 61
Lowman & Hanford Company, 14

MacReady, Lt. John A., 186
Male, Molly, 8
Mardon, Cora, 45, 51, 52, 53, 73, 80, 83, 84, 231n17; 1910 activities, 60; AYPE exhibit, 41; married, 28; sent home, 63
Mardon, William Bruce, 45, 46, 53, 60, 66, 68, 71, 74, 90; 1910 activities, 60; arrested, 33; AYPE exhibit, 41; death, 96; final parting, 94; importance to expedition, 28; joined Meeker, 27; married, 28; surgery, 65; threatened Ezra, 95
Marquis, John Abner, 197
Martin, George W., 73, 74
Maue, Daniel R., 201
Mayors: Carter H. Harrison (Chicago), 86; Charles W. Davison (San Jose), 51; E. H. Dewey (Nampa),

Index 263

68; Edwin M. Capps (San
Diego), 138; Frank K. Mott
(Oakland), 50; George Marshall
(Columbus), 81; George Ralph
(San Francisco), 114; Hiram H.
Edgerton (Rochester), 83; Joseph
Simon (Portland), 45; Joshua
F. Elder (Keokuk), 87; Samuel
Hays (Boise), 130; William B.
Thompson (Toledo), 83
McAuley, Thomas, 18
McClanahan, E. J., 204
McClellan Pass (Chinook Pass), 154
McCormack, Robert L., 58
McDonald, Roderick (Meeker's son-in-law), 6, 152
McDonald, Wilfred (Meeker's grandson), 6, 152
McGowan, F. C., 195
McKinnon, John Archibald, 163
Meacham 1906 monument: defaced, 64
Meacham, Walter, 171, 172, 203; created Old Oregon Trail Association (OOTA), 171
Meany, Edmund, 19, 215; Meeker conflict, 16
Medicine Creek Treaty, 12, 18
Meehan, William, 172
Meeker, Aaron (Meeker's half-brother), 41, 179
Meeker, Clara (Meeker's daughter-in-law), 41, 116
Meeker, Eliza Jane, 4, 10, 30, 31; death, 47; funeral, 47; in sanitarium, 44; lack of correspondence and separations from Ezra, 29
Meeker, Ezra: chicken business, 6; first lecture, 19; holly business, 6, 13, 52, 53, 110; phonographic recording, 101; president, WSHS, 7; radio, 177, 189
Meeker, Ezra (airplanes), 184, 244n32; 1910 aviation meet, 55; 1924 campaign flights, 163; 1924 cross-country flight, 186-188
Meeker, Ezra (disputes): AYPE, 45; Carrie Meeker, 84; Columbus

mayor, 81; Edmund Meany, 16;
illegally rented Mansion, 6, 12; Joe
Templeton, 85; Lewis and Clark
Exposition, 20; OTMA, 206, 212;
Panama Pacific Expo, 115; Robert
Bruce, 207; Willis Calder, 68;
WSHS, 8
Meeker, Ezra (expositions and expeditions): 1923 itinerary, 247n28; Klondike, 4, 6; OOTA pageant, 171; Second Oregon Trail Expedition, lecturing business, 79; Second Oregon Trail Expedition, purpose, 53
Meeker, Ezra (health issues): death, 216; hospitalized, 109, 148, 179, 214; leg injury, 113
Meeker, Ezra (motion pictures): covered wagon, 175; entered business 1911-12, 97; incorporated film company, 176; Los Angeles effort, 180; Naches Cliff scene, 161; purpose of 1923 trip, 177; purpose of *Kate Mulhall*, 191; reason for joining 101 Ranch, 190; Rose Parade film, 114; Yellowstone film, 114
Meeker, Ezra (Naches Pass): 1919 backpacking trip, 155; 1920 horse trip, 156; 1923 motion picture plan, 161; 1924 state legislature campaign, 162; Chinook Pass trip, 163; election defeat, 164; lost watch, 157; made highway *cause célèbre*, 153
Meeker, Ezra (OTMA): 1926 Oregon Trail trip, 201; annual meeting, 207; expanded coin idea, 196; Fort Hall article, 195; lobbied congress, 198; New England tour, 213; OTMA birth, 197; oxmobile, 213; passage of coin bill, 199; plan to replace board, 209; selling coins, 203; Senate testimony, 198
Meeker, Ezra (personal): birthdays, 54, 91, 116, 121, 170, 180; comments on wife's death, 47; expectations

regarding relatives, 84; gifted
  papers to WSHS, 117; guilt
  regarding wife, 31; separations
  from wife, 29; statue dedication,
  204; voting record, 215
Meeker, Ezra (predictions):
  automobile's future, 78; interstate
  bus service, 141; war with Japan,
  122
Meeker, Ezra (U.S. Presidents):
  Benjamin Harrison, 238n7; Calvin
  Coolidge, 189; Herbert Hoover,
  238n7; Theodore Roosevelt, 9, 33;
  Woodrow Wilson, 122
Meeker, Ezra (World War 1):
  food production, 149-151;
  grandchildren's role, 150; Herbert
  Hoover, 150; League of Nations,
  152; Victory Garden, 150, 151
Meeker, Fanny (Meeker's sister-in-law),
  48
Meeker, Frank (Meeker's nephew), 71
Meeker, Fred (Meeker's son), 4
Meeker, Grace (Meeker's
  granddaughter), 138
Meeker, Jacob (Meeker's cousin), 92
Meeker, Jessie (Meeker's niece), 48
Meeker, John (Meeker's brother), 10,
  62, 77; AYPE exhibit, 41; death,
  79
Meeker, Marion (Meeker's son), 4, 5,
  47, 52, 138, 152; driving ability,
  137; partner chicken business, 6
Meeker, Nathan, 71
Meeker, Olive (Meeker's daughter), 4,
  6, 47
Meeker, Rozene, 72
Mellon, Treasury Secretary Andrew
  W., 198
Melodeon, 26
Metropolitan Press Company, 226n24;
  printing *Pioneer Reminiscences*, 13
Meyers, Henry W., 107
Miller brothers: Joe, George and Zach,
  101 Ranch, 190-191
Miller, David B., 60

Miller, Joaquin (poet), 50, 137
Miller, U.S. Representative John, 198
Miner, Cora. *See* Mardon, Cora
Miner, Ethel, 95, 110
Moody, Malcom A., 69, 232n36
Moore, Wilbur B., 188
Moorhouse, Lee, 26
Motion pictures: *Covered Wagon*
  by Paramount, 176; Meeker's
  1911-12 efforts, 97; New York
  1922 events, 173; *Oregon Trail* by
  Universal, 176; realization of value,
  79; Rose Parade film, 113; today's
  location, 100
Murphy, Mabel, 195

Naches Pass (today), 165
Naches Pass 1853 wagon road, 154,
  155; apocryphal story, 153;
  Longmire-Biles wagon train, 153
Naches Pass highway: Chinook Pass
  competition, 154
Nadeau, Ira A., 39
National Highways Association, 33;
  Meeker Headquarters, 174
Nickerson, Hermon, 239n31

Ogle, Van, 155
Old Oregon Trail Association
  (OOTA): 1922 Baker City
  pageant, 171; created, 171
Old Oregon Trail Monument
  Expedition, 18, 23, 24;
  accomplishments, 31, 57; goals, 24;
  origins, 23
Oregon Trail map, 68, 73, 78, 93, 113;
  first modern effort, 58; intended
  for Congress, 66; on Pathfinder,
  124
Oregon Trail memorial coin: fraud, 210
Osborn, Chase, 198
Osborne, Carrie (Meeker's daughter),
  6, 10, 38, 50, 110, 139, 148;
  cared for mother, 30; cosigned
  loan, 44; dispute with father, 84;
  involvement in OTMA take over,

209; purchased Mansion, 5
Osborne, Eben (Meeker's son-in-law), 62, 84, 93, 110; death, 171; purchased Mansion, 5; reviewed book contract, 167; role in *Pioneer Reminiscences*, 13
Osborne, Olive (Meeker's granddaughter), 59, 151; married, 151
OTMA: goals, 250n5; membership, 198; origin, 112; Yale monument, 210
Otter Massacre visit, 69
Ox Dandy, 27, 89, 107, 112, 113, 118; AYPE exhibit, 41; death, 109; joined expedition, 27
Ox Dave, 62, 89, 107, 112, 118; AYPE exhibit, 41; joined expedition, 27
Ox Dick, 88, 89, 91, 92
Ox Harry, 88, 89
Ox Stub, 21, 27
Ox Twist, 21, 27; death, 27
Oxen: Smithsonian donation plan, 90, 91, 117
Oxmobile, 213, 214; fate of, 253n9

Packard, Lulu, 20
Palmer, Jesse, 74
Panama Pacific Exposition 1915, 111
Pathfinder Expedition, 119, 120; concerns about driver, 121; driver #2, Earl S. Shaffer, 124; driver #3, Charles J. Wyley, 125; driver #4, John Mattlock, 125; driver #5, Ernest Grot, 126; driver #6, Marion Meeker, 137; driver #7, Edward and William Johnstone, 140; driver #8, Dan Lang, 142
Pathfinder Expedition Part 1: company complaints, 126; in Yellowstone Park, 129; Lincoln Highway competition, 133; mechanical problems, 132; Meeker's inability to drive, 122; paying driver, 127; company praise, 132
Pathfinder Expedition Part 2: mechanical problems, 136; mission in California, 136
Pathfinder Expedition Part 3: fate of the Pathfinder, 148; Meeker hospitalized, 148
Paulhamus, William H., 44
Phy, Dr. William T., 64
Piles, Senator Samuel, 54, 64, 76, 77, 78, 236n42
Pioneers of America film corporation, 197; incorporated, 176
Pratt, George Dupont, 112, 166, 173; Meeker on retainer, 173; withdrew film project funding, 180
Price, Jacob, 43, 46
Puyallup street names controversy, 105

Rainey, Joe, 130
Rogers, Palmer, 195, 197
Roosevelt, President Theodore, 27, 66, 96, 107; audience with Meeker, 34
Rose Parade, 36, 54, 113
Ross, Charles, 183, 216

San Francisco's Good Roads Parade, 136
San Francisco's Portola Day Parade, 46, 49
Sargent, Nelson, 10
Schwartz, Rose Jay, 214
Scott, Harvey, 19, 58, 59
Scott, William W., 100
Second Expedition: failures, 57; markers, 60, 61, 63, 64; plan, 58; successes, 58
Seymour, Edmund, 197; Meeker dispute, 208
Shakleford, Mrs. Roxie, 26
Simmons, Michael, 113
Simonds, W. D., 46, 47, 229n26
Smith, Reverend Lincoln, 216
Smith, U.S. Representative Addison, 198
Snoqualmie Pass, 154
Sorger, Ed, 27
Spinning, Frank, 183

Standrod, Mrs. Drew W., 70, 129
Steamboat *Constitution*: 1858 trip, 184
Steamboat *Northwest*, 19
Stereopticon, 26
Steen, Ralph, 214
Stevens, Hazard, 12, 14
Studebaker, John Mohler, 121
Swan, Sam, 172

Templeton, Bertha (Meeker's granddaughter), 41, 116, 150
Templeton, Dr. Charles (Meeker's grandson), 110, 179
Templeton, Ella (Meeker's daughter), 46, 152, 215
Templeton, Joe (Meeker's grandson), 50, 55, 85; acerbic exchange with grandfather, 86; cosigned loan, 44; dispute with Ezra, 85; involvement in OTMA take over, 209
Templeton, Reverend Harry (Meeker's grandson), 47, 152, 216

The Days of Ezra Meeker Corporation, 183
Thompson, Reginald H., 60
Tittle, William S., 138
Turner, Joseph R., 176, 213

U.S. Army Havilland-DH 4 airplane, 187

Warner, Arthur B., 8
Webber, Bert, 210
Weir, Allen, 9
Whitman, Marcus, 21
Wilson, President Woodrow, 112, 147; audience with Meeker, 122
Wilson, Robert (Meeker's tax accountant), 10, 11
Wright, Orville, 188, 23n8
Wyllys, Charles, 36

Acknowledgments

On July 19, 1999, my wife Pat and I stood in front of the Meeker monument at Emigrant Springs State Park in Oregon's Blue Mountains. She wondered aloud where the rest of the monuments were. Two decades later, with her invaluable help, my expanded Meeker quest has come to an end. It has been quite a journey.

I would like to thank Mary Lein of the Pocatello Public Library for connecting me with Mike Fagerquist, who connected me with Jack Evans of La Grande, Oregon, who introduced me to the Oregon-California Trails Association. I also would like to acknowledge Gary Greenbaum who gave me much insight into the story behind the Oregon Trail Memorial half-dollar coin and who proofread some of the chapters.

Many thanks also go out to the members of the Puyallup Historical Society at the Meeker Mansion. In particular I wish to thank the Andersons, the Egans, and the Perkinsons.